The Birth of Modern Science

The Making of Europe

Series Editor: Jacques Le Goff

The Making of Europe series is the result of a unique collaboration between five European publishers – Beck in Germany, Blackwell in Great Britain and the United States, Critica in Spain, Laterza in Italy and le Seuil in France. Each book will be published in all five languages. The scope of the series is broad, encompassing the history of ideas as well as of societies, nations, and states to produce informative, readable, and provocative treatments of central themes in the history of the European peoples and their cultures.

Also available in this series

The European City
Leonardo Benevolo

The Rise of Western Christendom: Triumph and Diversity 200–1000 AD
Peter Brown

The European Renaissance
Peter Burke

The Search for the Perfect Language
Umberto Eco

The Distorted Past: A Reinterpretation of Europe
Josep Fontana

The European Family
Jack Goody

The Origins of European Individualism
Aaron Gurevich

The Enlightenment
Ulrich Im Hof

The Population of Europe
Massimo Livi-Bacci

Europe and the Sea
Michel Mollat du Jourdin

The Culture of Food
Massimo Montanari

The First European Revolution, 900–1200
R. I. Moore

Religion and Society in Modern Europe
Réne Rémond

The Peasantry of Europe
Werner Rösener

The Birth of Modern Science
Paolo Rossi

States, Nations and Nationalism
Hagen Schulze

European Revolutions 1492–1992
Charles Tilly

In Preparation

Democracy in European History
Maurice Agulhon

Migration in European History
Klaus Bade

Women in European History
Gisela Bock

Europe and Islam
Franco Cardini

The Industrialization of Europe
Jurgen Kocka

The Law in European History
Peter Landau

The Frontier in European History
Krzysztof Pomian

The Birth of Modern Science

Paolo Rossi

Translated by
Cynthia De Nardi Ipsen

BLACKWELL
Publishers

Copyright © Paolo Rossi 2000
English translation copyright © Blackwell Publishers Ltd 2001

First published in Italian as *La nascita della scienza moderna in Europa* by Gius.
Laterza & Figli in 2000.
First published in English by Blackwell Publishers Ltd 2001, and by three other
publishers: © 2001 Beck, Munich (German); © 2001 Critica, Barcelona (Spanish);
© 2001 Editions du Seuil, Paris (French)

2 4 6 8 10 9 7 5 3 1

Blackwell Publishers Ltd
108 Cowley Road
Oxford OX4 1JF
UK

Blackwell Publishers Inc.
350 Main Street
Malden, Massachusetts 02148
USA

British Library Cataloguing in Publication Data

A CIP catalogue record for this book is available from the British Library.

Library of Congress Cataloging-in-Publication Data

Rossi, Paolo, 1923–
    [Nascita della scienza moderna in Europa. English.]
    The birth of modern science / Paolo Rossi ; translated by Cynthia De Nardi Ipsen,
    Carl Ibsen.
        p.   cm. — (Making of Europe)
    Includes bibliographical references and index.
    ISBN 0–631–20562–4 (alk. paper) — ISBN 0–631–22711–3 (alk. paper)
    1. Science—Europe—History.   I. Title.   II. Series.

    Q127.E8 R6713 2001
    509.4—dc21
                                                                    00–049429

Typeset in 10.5 on 12 pt Sabon
by Ace Filmsetting Ltd, Frome, Somerset
Printed in Great Britain by T.J. International, Padstow, Cornwall

This book is printed on acid-free paper

Contents

Series Editor's Preface, by Jacques Le Goff   viii

Introduction   1
European Science – A Revolution and its History –
About this Book

1   Obstacles   9
Forgetting What we Know – Physics – Cosmology – Vile
Mechanics

2   Secrets   18
"Margaritae ad porcos" – Hermetic Knowledge – Public
Knowledge – The Hermetic Tradition and the Scientific
Revolution – Secrets and Public Knowledge

3   Engineers   29
Practice and Discourse – Engineers and Machines – The
Workshop – Leonardo da Vinci – "Discourse" and
"Construction" – Knowledge that can Grow – Art and
Nature – Daedalus and the Labyrinth

4   The Unseen World   41
Printing – Ancient Books – The Old and the New –
Illustration – New Stars – The Discovery of an Unknown
Visual World – The New World

5   A New Universe   56
Copernicus – The World in Pieces – Tycho Brahe – Kepler

6   Galileo   73
    Early Works – Astronomical Discoveries – Nature and
    the Bible – Hypotheses and Realism – Copernicus is
    Condemned – The Book of Nature – The *Chief
    Systems* – The Destruction of Aristotelian Cosmology –
    Geometrization, Relativity, and Inertia – The Tides –
    Galileo's Tragedy – The New Physics

7   Descartes   99
    A System – "I Come Forward Masked" – Mathematical
    Equations meet Geometry – Physics and Cosmology –
    The World as Geometry

8   Countless Other Worlds   108
    An Infinite Void – An Infinitely Populated Infinite
    Universe – Galileo, Descartes, and the Infinite Universe
    – We are not Alone in the Universe – Huygens'
    Conjectures – The End of Anthropocentrism

9   Mechanical Philosophy   122
    The Need for Imagination – Mechanisms and
    Mechanics – The Natural and the Artificial: To Know
    and to Make – Animals, Men, and Machines – Can
    Mechanists still be Christians? – Leibniz and the Critique
    of Mechanics

10 Chemical Philosophy   139
    Chemistry and its Forefathers – Paracelsus –
    Paracelsians – The Iatrochemists – Chemistry and
    Mechanical Philosophy – Mechanics and Vitalism

11 Magnetic Philosophy   148
    Strange Events – Gilbert – Jesuits and Magic – Cautious
    Experiments and Daring Devices – The Sulfur Globe –
    Music and Tarantism

12 The Heart and Generation   157
    The Sun of the Microcosm – Ovists and Animalculists –
    Preformation Theory

13 Time and Nature   164
    The Discovery of Time – Strange Rocks – How are
    Natural Objects Made? – A Sacred Theory of the Earth

– Leibniz and the *Protogaea* – Newtonians and
Cartesians

14 **Classification**   175
"Poa bulbosa" – Classification – Universal Languages –
A Language for the Discussion of Nature – To Name is
to Know – Mnemonic Aids – The Essential and the
Incidental

15 **Instruments and Theories**   183
Sensory Aids – Intellectual Aids

16 **Academics**   192
The University – The Academy – The First Academies –
Paris – London – Berlin – Bologna – Publications

17 **Newton**   203
The *Principia* – The "General Scholium" – The *Opticks*
– Newton's Life – A Brief Interlude on the
Manuscripts – The "Queries" – The Cosmic Cycles –
Chronology – Knowledge of the Ancients – Alchemy –
Newton's Religion and the Apocalypse – Interpreting
the Bible and Nature – Conclusion

Chronology   230
Bibliography   235
Index   251

Series Editor's Preface

Europe is in the making. This is both a great challenge and one that can be met only by taking the past into account – a Europe without history would be orphaned and unhappy. Yesterday conditions today; today's actions will be felt tomorrow. The memory of the past should not paralyze the present: when based on understanding it can help us to forge new friendships, and guide us towards progress.

Europe is bordered by the Atlantic, Asia, and Africa, its history and geography inextricably entwined, and its past comprehensible only within the context of the world at large. The territory retains the name given it by the ancient Greeks, and the roots of its heritage may be traced far into prehistory. It is on this foundation – rich and creative, united yet diverse – that Europe's future will be built.

The Making of Europe is the joint initiative of five publishers of different languages and nationalities: Beck in Munich; Blackwell in Oxford; Critica in Barcelona; Laterza in Rome; and le Seuil in Paris. Its aim is to describe the evolution of Europe, presenting the triumphs but not concealing the difficulties. In their efforts to achieve accord and unity the nations of Europe have faced discord, division, and conflict. It is no purpose of this series to conceal these problems: those committed to the European enterprise will not succeed if their view of the future is unencumbered by an understanding of the past.

The title of the series is thus an active one: the time is yet to come when a synthetic history of Europe will be possible. The books we shall publish will be the work of leading historians, by no means all European. They will address crucial aspects of European history in every field – political, economic, social, religious, and cultural. They will draw on that long historiographical tradition which stretches back to Herodotus, as well as on those conceptions and ideas which have transformed historical enquiry in the recent decades of the twentieth century. They will write readably for a wide public.

Our aim is to consider the key questions confronting those involved in Europe's making, and at the same time to satisfy the curiosity of the world at large: in short, who are the Europeans? where have they come from? whither are they bound?

Jacques Le Goff

Introduction

European Science

That complex historical reality which we today call *modern science* was not born in any one specific place in Europe. Its birthplace is all of Europe. We should remember that Copernicus was Polish; Bacon, Harvey, and Newton English; Descartes, Fermat and Pascal French; Tycho Brahe Danish; Paracelsus, Kepler, and Leibniz were Germans; Huygens Dutch, and Galileo, Torricelli, and Malpighi were Italians. The arguments of one were linked to those of another in an artificial or imaginary reality without borders; in a Republic of Science that worked itself, against the odds, into difficult, often dramatic, and sometimes even tragic social and political contexts.

Modern science was not established in the serenity of a college campus or in the artificial environment of a research laboratory: places near to but not immersed in the dramatic flow of historical events. And this was so for one reason alone: such institutions (at least those relating to "scientific" knowledge) did not yet exist. The ivory towers of the "natural philosophers" – so productive and yet so unjustly maligned in our time – had not yet been constructed.

Though almost all scientists in the seventeenth century studied at universities, very few continued their careers there. The university was *not* the center of scientific research. Modern science was born outside the academy and frequently in opposition to it, and over the course of the seventeenth century – and especially during the eighteenth and nineteenth centuries – it grew into an organized social enterprise capable of spawning its own institutions.

The tumultuous behind-the-scenes events in the history of physics, astronomy, and chemistry are rarely mentioned in scientific textbooks. Those events occurred in a specific historical context, and I urge the reader of this book (which is about ideas, theories, and experiments and out of necessity can concede little of their telling) to keep that context in mind. The times in which the so-called "founding fathers" of modern science lived did not only coincide with the music of Monteverdi and Bach, the theater of Corneille and Molière, the artwork of Caravaggio and Rembrandt, with Borromini's architecture and Milton's poetry. The 160 years between *De revolutionibus* (1543) by Copernicus and Newton's *Opticks* (1704) were critical ones in the continent's dramatic history, and Europe was a radically different place to live in then (even in terms of everyday life) than it is today.

In the winter of 1615–16, six witches were burnt at the stake in the small Swabish town of Leonberg. In nearby Weil-der-Stadt, a small village of 200 homes, 38 witches were burnt at the stake between 1615 and 1629. In Leonberg, the wife of a glazier made several accusations against a peculiar old woman named Katharine: she had given a neighbor a magic drink which made her ill; she had cast the evil eye on a tailor's children who then died; and she had paid the grave-digger for her father's skull in order to have it made into a goblet as a gift for one her sons, an astrologer and student of black magic. One day, a twelve-year-old girl on her way to the kiln with raw bricks passed the old woman on the road and was seized by such terrible pains in the arm and hand that for several days she could not move hand or fingers. To this day, lumbago and a stiff-neck are called *Hexenschuss* in German, *Hekseskud* in Danish, and *colpo della strega* in Italian. The 73-year-old woman was accused of witchcraft and held in chains for months. She was indicted on 49 counts, threatened with torture, and brought before the executioner who treated her to a detailed description of the many tools he used in his trade. Katharine was finally acquitted on October 4, 1621, after more than a year in prison and six years after she was first accused of witchcraft. She could not return to Leonberg as the local population would have lynched her (Caspar, 1962: 249–65).

The old woman had a famous son named Johannes Kepler who worked feverishly in her defense. During the long years of her trial he not only produced hundreds of pages to save his mother from torture and death, but also wrote the *Harmony of the World*, a part of which would later come to be known as Kepler's Third Law. It was Kepler's belief that celestial harmony reigned at the center of the world, "like a sun shining through the clouds." He knew, however, from experience, that a similar harmony did not reign on earth. In the sixth chapter of his book on planetary sounds, Kepler observed that since the Earth produced the notes "Mi-Fa-Mi" it was reasonable to conclude that *MI*sery and *FA*mine ruled that planet. He finished writing this book three months after the death of his daughter, also named Katharine.

Very few scientists of the day quietly dedicated their lives to research. Nor need we recall the more dramatic examples like Giordano Bruno's burning at the stake or Galileo's trial for heresy; one gets a good sense of these times simply by reading Adrien Baillet's *The Life of Descartes*. They were decades not just of witchcraft trials and the Inquisition but also of the Thirty Years' War, the literal meaning of which we rarely stop to consider. During those years Europe was criss-crossed by armies of mercenaries (and following in their wake a stream of artisans, cooks, prostitutes, runaways, and traveling salesmen) committing theft, fraud, arson, rape and murder, and destroying harvests, desecrating churches and sacking villages. In the same period, the populations of European cities like Milan, Seville, Naples, and London were being halved by the plague, which had all the features of a long and terrible epidemic. The events described by Defoe and Manzoni for London and Milan occurred not once, but over and over again.

It could only have been in the context of an ideal Republic, independent of strife, discord, and earthly miseries, that Francis Bacon could have made the astonishing statement that science for glory or national power was morally less noble than science for the good of all mankind. Or that Father Marin Mersenne in referring to Canadian Indians and Western peasants could have stated that "there is nothing that one man can do that cannot be done by another and every man possesses all that is necessary for philosophizing about and discussing all things" (Mersenne, 1634: 135–6). The protagonists of the Scientific Revolution had, moreover, something compelling in common: the consciousness that something was being *created* through their work. The term *novus* or new recurs almost obsessively in hundreds of titles of seventeenth-century scientific works: from *Nova de universis philosophia* by Francesco Patrizi and *Newe Attractive* by Robert Norman to Bacon's *Novum Organum*, and even Kepler's *Astronomia Nova* and Galileo's *Discourses on Two New Sciences*.

A form of knowledge that differed structurally from other cultural forms took root and matured in those years, and laboriously created its own institutions and lexicon. This knowledge required "judicious experiment" and "irrefutable proof" and, for the first time, demanded that the two complicated things *go together*, that they be inextricably intertwined. Declarations had to be "public" or subject to confirmation by others; they had to be presented and demonstrated to others, discussed and subjected to possible refutation. In that milieu some admitted to being wrong, to not being able to prove what they had set out to, and bowed to the proofs of others. Clearly this did not happen often, for resistance to change was very strong (as it is for all human groups), yet the very fact that in that moment it was established that the truth of a proposition was independent of the authority of its author and was in no way bound to *revelation* or *illumination* contributed to an ideal legacy that Europeans still today regard as an unalienable right.

A Revolution and its History

The birth of modern science has been, and is still, (justifiably) referred to as a "Scientific Revolution." Revolutions do not just look to the future and give rise to something which previously did not exist, but they also construct an imaginary, and often negative, past. For instance, the introduction of the Enlightenment *Encyclopédie,* or even the beginning of Rousseau's *Discourse on Science and Art*, reveal the great tendency in the late eighteenth century to define the Middle Ages as a dark age, a "relapse into an uncivilized state" that was ended by the splendor of the Renaissance.

In principle, historians do not accept "imaginary pasts." They even debate the attempts of individuals to locate themselves within the historical process. The one thousand years of history generically referred to as the "Middle Ages," during which quite a few great intellectual revolutions took place, have been meticulously studied since the middle of the nineteenth century. We know today that the myth of the Middle Ages as a barbaric era was precisely that, a myth constructed by humanists and the fathers of modernity. Over the course of the Middle Ages there were many wonderful churches, cathedrals, convents, and windmills built; fields were tilled by the heavy plow; the stirrup was invented and this changed the nature of warfare and European politics by transforming the mythical classical Centaur into a feudal Lord (White, 1962).

Cities were becoming more and more populated, and became the centers of both commercial and intellectual exchange. We associate great medieval philosophy with the meeting of different cultural traditions: Christian, Byzantine, Hebrew, and Arab (De Libera, 1991). This was the milieu that created the university and established the figure of the *intellectual*, an individual who, between the twelfth and thirteenth centuries, came to be recognized as exercising a particular profession and whose specific task it was to produce and transmit the "liberal arts" (Le Goff, 1993). Universities were founded in Bologna, Paris, and Oxford at the end of the twelfth century, grew in number over the next hundred years, and spread throughout Europe in the fourteenth and fifteenth centuries. The university became the privileged site of a kind of knowledge that was considered worthy of social status and remuneration; a brand of knowledge with its own meticulously established laws (Le Goff, 1977: 153–70). Unlike the schools of monasteries or cathedrals, the university was a *studium generale* and had a precise legal status established by a "universal" authority (such as the Pope or an Emperor). The fact that teachers could teach anywhere (*licentia ubique docendi*) and students moved freely contributed in great measure to building the unity of Roman Catholic culture. The use of Latin as the language of scholarly communication transformed medieval universities into international study centers where men and ideas circulated freely (Bianchi, 1997: 27). The "scholastic method" (based on *lectio, quaestio, disputatio*) left an indelible mark on European culture, and it is quite true that in order to understand many modern philosophers, beginning with Descartes, it is absolutely necessary to go back to the work of those authors they so fervently opposed.

Medieval philosophy and science have been widely studied beyond simply the topics of the secularization of culture and the Church's condemnations of many philosophical theories. A number of scholars, for example, support the notion of a strong continuity between the science of Merton College scholars (such as Bradwardine), the "Parisian physicists" (such as Nicole Oresme and Jean Buridan), and the work of Galileo, Descartes, and Newton. Space prevents me from discussing the interpretations of Pierre Duhem (Duhem, 1914–58) or Marshall Clagett (Clagett, 1959) in these pages, so let me simply list some of the good reasons for believing the opposite argument, namely that there was a strong discontinuity between the medieval scientific tradition and modern science, a discontinuity that fully justifies use of the expression "Scientific Revolution."

1 Modern scientists and medieval philosophers had radically different views of nature. The modern scientist, unlike his predecessor, sees no *essential* difference between a natural body and an artificial one.
2 Modern scientists conduct their inquiry of nature under artificial conditions: Aristotelians discussed experience in terms of the everyday world to illustrate or exemplify a theory while the "experiences" of modern scientists are artificially created *experiments* aimed at confirming or falsifying theories.
3 Modern scientific knowledge resembles the exploration of a new continent, while that of the medieval thinker was more like a patient probing of problems according to codified rules.
4 As compared to modern scientific criticisms, Scholastic knowledge failed to interrogate nature and only questioned itself, and in such a way as to *always* provide a satisfactory answer. In that tradition there was room for the master and the disciple but not for the inventor.
5 The modern scientist, above all Galileo, worked with a "freedom" and "methodological opportunism" that was unknown in the medieval tradition (Rossi, 1989: 111–13). The medieval claim to absolute precision hindered rather than fostered the creation of a

mathematical science of nature. Galileo continued inventing ever-more-precise systems of measurement but "shifted the focus from ideal precision to that precision required for the goal at hand, and one achievable by means of the available instruments [. . .] the paralyzing myth of absolute precision was one of the reasons that fourteenth-century thinkers were not able to progress from abstract *calculationes* to an actual quantitative study of natural phenomena" (Bianchi, 1990: 150).

Still, it is not so much this short list that will justify my insistence on referring to modern science as an intellectual revolution, but the pages that follow.

About this Book

I was honored to be asked by Jacques Le Goff to write a book called *The Birth of Modern Science in Europe*. The European publishers justifiably imposed a strict limit of 85,000 words or 300 pages of 1,800 characters each. I have exceeded them, though not by much.

Just a list of everyone we call a *scientist* (to use a nineteenth-century term) who lived between the birth of Copernicus and the death of Newton, and who is worthy of mention in a history of science textbook, would fill pages and pages. If, in addition, I were to list one or two of their most important contributions, the situation would become out-of-hand.

So I immediately abandoned any pretext of completeness, and as a result also gave up on writing a *textbook* of the history of science. Moreover, I would like to share with the reader some of the choices I have made in order to alert you to what can be found in this book, and also clarify my point-of-view.

This book covers topics such as the new astronomy, discoveries made with the aid of telescopes and microscopes, the principle of inertia, experiments on voids, the circulatory system, the great discoveries of *calculus*, and so forth, but each chapter also addresses the great ideas and themes that were central over the course of the "revolution": the rejection of a *Hermetic* concept of knowledge, the new appraisal of technology, the hypothetical or realistic character of our knowledge of the world, the attempt to apply – in part even to the world of man – the models of *mechanical philosophy*, a new view of God as an engineer or clockmaker, and the introduction of the dimension of time into the study of nature.

With regard to method, I believe that the specific theories that are the *kernel* of any science are not actually the reflection of specific socio-historical conditions. Instead, and this has been the direction of all my work up to this point, I am certain that *history* has much to do with our cultural *images of science*, or the discussion of what science *is* and *should be*. In many cases those images have influenced the acceptance or success of a particular theory. A certain image of science can often be used to define the *frontiers* of science, or to distinguish science from magic, metaphysics, or religion. It also can be the basis upon which a problem is chosen, from an unlimited number of them, to be solved.

That which today seems firmly codified and transmitted by a biology and physics textbook – what seems obvious and natural – is really the outcome of choices, options, conflicts, and alternatives. The alternatives and the choices, *before* their eventual codification, were real and not imaginary. Every choice involved options, conflicts and rejections, and was at times made in a most dramatic way.

I hope that this book makes the following points clear: that the philosophy of historical continuity is misleading and artificially imposed upon the actual historical process; that historical research never unearths monoparadigmatic stages or typical periods in time; that critical dialogue between theories, scientific traditions, and images of science has always been (and continues to be) continuous and unceasing; that seventeenth-century science was, at one and the same time, Paracelsian, Cartesian, Baconian, and Leibnizian; that non-mechanical models were strongly operative even in unexpected places; that the emergence of problems and potential fields of inquiry was deeply linked to philosophical and metaphysical discussions; and that the figure of the *scientist* emerged at different times and in different ways in the individual scientific disciplines, given that in some cases (such as mathematics and astronomy) ancient traditions were revived, while in other cases specific ancient traditions were singled out, and in still others it was vital for cognitive and experimental work to have a new or "alternative" nature.

Historians must constantly remind not only their readers but also their fellow intellectuals, philosophers and scientists of one seemingly obvious thing, something that takes constant reminding because everyone (and so even the most refined philosopher and scientist) has an almost unsquelchable tendency to forget it: all those who worked on, thought about, formulated theories and conducted experiments at the time of the *birth* of modern science lived in a world so very different from our own, one in which views that today seem culturally irreconcilable co-existed. Interest in alchemy, for example, flourished in the seventeenth century at the same time as there was an extraordinary burst of mathematical creativity. Newton was one of the great creators of infinitesimal calculus but also wrote over a million words on the topic of alchemy (roughly ten times longer than this book). Seventeenth-century scientists could not know what is clear to us today, "that seventeenth century alchemy was the last blossom from a dying plant and seventeenth century mathematics the first blooming of a hardy perennial" (Westfall, 1980: 290, 23).

I am certain that what we call "science" acquired in those years some of the basic characteristics which still distinguish it today and which its founders correctly believed to be new in the history of mankind: an expanding project or collective enterprise dedicated to knowing and exploring the world. An enterprise which certainly was not, and never claimed to be, innocent, and unlike political, artistic, religious, and philosophical ideals, became a powerful *unifying force* in the history of the world.

This book was not written for the historian or philosopher of science. It is written for those young people seeking to establish a personal relationship with the history of ideas and the complex, prolific, fascinating things which are science and philosophy. I especially had in mind as an audience the many people (among whom I count a number of dear friends) who have dedicated themselves to "humanistic" studies, who think of science as something "dry" and are firmly convinced that it is at best nominally relevant to culture and the history of culture, whose image of science and its history is a conveniently reductive one which many twentieth-century philosophers (often illustrious ones) have contributed to and promoted, and who – usually unbeknownst to them – agree with the early twentieth-century notion of science as somehow *bankrupt*.

Since the pages to follow represent an attempt to synthesize (and also revise) forty years worth of work on ideas relating to the Scientific Revolution, a list of acknowledgments would be far too lengthy, to the many friends and many young, and now not-so-young, scholars. Instead, I dedicate this book to my dear, determined and unexpected granddaughter Giorgia, who has the same enchanting blue eyes as her grandmother Andreina.

1

Obstacles

Forgetting What we Know

Historians are interested in the different ways in which the human brain has worked at different times in history more than in its permanent structure. In order to understand different ways of thinking we must try to forget what we already know or think we know. Often we must even follow the reasoning or adopt the metaphysical principles held by past thinkers, for they were based as much on logic and inquiry as modern principles of mathematical physics and astronomy (Koyré, 1971: 77). In the words of Thomas Kuhn, it is essential to try to *unlearn* "the thought patterns induced by experience and prior training" (Kuhn, 1980: 183).

In the 1930s, the French philosopher Gaston Bachelard coined the term *epistemological obstacle* to refer to those convictions – drawn from both general and scientific knowledge – that tend to block breaks and discontinuities in the development of scientific knowledge and so function as powerful obstacles to the assertion of new truths. The nature of Bachelard's inquiry reinvigorated the historical study of science and transformed it from a "hallowed sequence of discoveries" to a history of the arduous paths of reason.

A specific example will help to illustrate several of Bachelard's central concepts. Those concepts include: (1) epistemological obstacles; (2) the distance between science and the realism of common sense; (3) a false notion of historical continuity deriving from language itself. Until the nineteenth-century invention of the light bulb it was an accepted truth that illumination was produced by the burning of some material. But Edison's glass bulb prevented the burning of a material; it was designed not to protect a flame but to create a vacuum for a filament. An ancient lamp and a modern one then had only one thing in common, and that only from the perspective of everyday life: their purpose was to defeat the dark. In addition, the technical development actually involved the complex *theory of combustion* which in turn involved the equally complex history of the discovery of oxygen (Bachelard, 1949: 104; Bachelard, 1938).

Physics

Students today have studied the difference between the *weight* of a body (which changes with its distance from the Earth) and its *mass* (which in classical or pre-Einsteinian physics is the same everywhere in the universe). They have also studied Newton's first law, the *principle of inertia*, according to which in the absence of an external force it takes the application of a force to stop a body that is moving in a straight line at a constant speed, and that constant, straight-line motion is for that matter, like rest, one of a body's "natural" states. And they are also familiar with Newton's second law, according to which *acceleration* and not velocity is proportional to the amount of force applied (unlike Aristotle's assertion that a given amount of force gives a body a determined velocity). In addition, these students also know something that was inconceivable according to ancient or pre-modern physics: a *constant* force gives a body a *variable* motion (uniformly accelerated) and any force, however small, is capable of this action on any body, however large. The modern student not only knows that every circular motion is an accelerated motion and that circular motion is not the prototype for perpetual planetary motion, but, unlike pre-Newtonian physics and Galileo's own beliefs, she knows that circular motion is in fact not "natural" but can be *explained* by a centripetal force that causes a body to deviate from the natural straight line path it would follow in the absence of that force.

The history of physics, from the late-Scholastic *impetus* theory to the crystal clear pages of Newton's *Principia*, has been the history of a profound conceptual revolution that radically altered notions of motion, mass, weight, inertia, gravity, force, and acceleration. At the same time it involved new methods and a revised overall concept of the physical universe and led to new ways of defining the ends, tasks, and goals of our knowledge of nature.

Deeply held convictions had to be broken in order to arrive at the formation of Galileo and Newton's "classical physics," and it was the apparent self-evidence of these convictions that posed a great obstacle to the foundation of modern science. Commonly-held ideas were plain not only because they were based on ancient and deeply-rooted traditional knowledge but also because they appealed strongly to common sense. In fact the three general convictions that follow, each rejected by modern science, do seem to be reasonable generalizations of empirical observations:

1 Bodies fall because they are heavy and tend to their *natural place* at the center of the universe. Motion is therefore intrinsic, and the heavier the body, the faster the fall. The velocity of the fall is directly proportional to weight: if a 1 kg. ball and a 2 kg. ball are dropped at the same time, the 2 kg. ball will land first and the 1 kg. ball will take twice the time to reach the ground.

2 A medium through which a body moves is an essential element in the phenomenon of movement, and must be taken into account when determining the velocity of falling objects. The velocity of a freely falling body (directly proportional to its weight) was generally considered to be inversely proportional to the density of the medium. In a void or vacuum (a space without density) motion would be instantaneous and velocity infinite: a body could be in several places at the same time. These were all strong arguments against the existence of voids.

3 Because everything that moves is moved by something else (*omne quod movetur ab alio movetur*), the violent motion of a body is produced by a force acting upon it. It is unnecessary to find an explanation for the lasting resting state of a body because rest is the natural state of bodies. Motion (any kind of motion, whether natural or violent) is an unnatural and temporary state (with the exception of the "perfect" circular motion of the heavens) and ceases as soon as force is no longer applied; it is faster the greater the force applied. Given a fixed applied force, a body moves more slowly the greater its weight and when that force is no longer applied, movement also ceases: *cessante causa, cessat effectus*, or in other words, when the horse stops so does the cart.

These generalizations were drawn from such everyday experiences as the fall of a feather and a stone, and the motion of a horse-drawn cart. They were also associated with an anthropomorphic view of the world in which the immediacy of human feelings, behaviors, and perceptions are used as criteria for reality. The "errors" in ancient physics were deeply-rooted in our physiology and psychology. Why, Descartes asks in the *Principles* (1644), as a rule do we believe that more action is required for motion than for rest? We have made this mistake, he writes, "from the beginning of our days" because we are used to moving our bodies according to our will and the body knows it is at rest only because "it is attached to the Earth by its weight which we do not feel." Because we feel resistance when we move our limbs and this causes fatigue "we think that it takes more force and action to produce than to stop movement" (Descartes, 1967: II, 88).

Modern science, on the other hand, does not rely on generalizations drawn from empirical observation but rather on *abstract* analysis that divorces itself from common sense, emotions, and the immediacy of experience. It was the application of mathematics to physics that made the conceptual revolution of physics possible, and Galileo, Pascal, Huygens, Newton, and Leibniz made important contributions to this end.

Cosmology

At this point let us review some of the basic aspects of the traditional *world view* that was dismantled by Copernicus, Tycho Brahe, Descartes, Kepler, and Galileo.

First of all, there were differences between the *celestial* and *terrestrial* world, and *natural* and *violent* motion. Aristotelian philosophy claimed that the terrestrial or sublunar word was made up of a combination of four basic elements: earth, water, air, and fire. The weight of a given body depended on the proportions of each element because earth and water naturally tended to descend while air and fire tended to ascend. Generation and alteration in the sublunar world stemmed from the stirring up and mixing of the elements. Heavy bodies naturally fall downward and light bodies naturally rise upward: rectilinear motion up or down (considered absolute and not relative) depended on a body's tendency to reach its natural or assigned place in nature. This was confirmed by everyday experiences like objects falling through the air, fire rising, and bubbles floating on water. Yet experience also constantly presents us with other kinds of movement: throwing a stone up in the air, shooting an arrow from a bow, a flame licking downward because of the wind. These were *violent motions*, caused by an external force that countered the nature of the object it acted upon. *Cessante causa, cessat effectus*: when the force ceases, the object returns to its natural place.

Movement in Aristotelian physics was conceptually different from motion in modern physics. Movement was generally defined as any transition from a potential to a realized state. For Aristotle "movement" included motion in space, the alteration of properties, and the generation and corruption of matter. "Movement" referred both to physical phenomena and to phenomena that we regard today as chemical or biological. Motion *was a process or a becoming but not a state*. A body in motion did not change only in relation to other bodies: insofar as it was moving, the body itself was subject to change. Motion was a sort of quality that affected a body.

The terrestrial world was one of change, of birth and death, of generation and corruption. The heavens, instead, were immutable and eternal; heavenly motions were regular, and nothing there was born or corrupted. The stars and the planets (including the sun) moving around the earth were not made up of the same elements that composed bodies in the sublunar world but rather of a fifth heavenly element: ether or *quinta essentia* which was solid, crystalline, weightless, transparent, and resistant to change. The sun, moon, and other planets were fixed (like "knots in a piece of wood") to the equator of these rotating heavenly bodies, also composed of the fifth element.

The opposite of the rectilinear, irregular, and finite motion of the terrestrial world was the circular, uniform, and continuous motion of the planets and heavenly bodies. Because circular motion was perfect it was ideally suited to the perfect nature of the heavens. Without beginning or end, it was not directed towards anything, it continuously returned to itself and proceeded indefinitely. Ether filled the entire universe except for the terrestrial (or *sublunar*) world. Limited by the sphere of the fixed stars, the universe was finite. The divine sphere or prime mover carried the fixed stars and produced the motion that was transmitted by contact to the other spheres and reached finally that of the moon, the lowest boundary of the celestial world. By definition, there was no circular motion on Earth, which stood fixed at the center of the universe. The centrality and immobility of the Earth was a theory not only plainly confirmed by everyday experience but also the basis of all Aristotelian physics.

The grandiose heavenly machine envisioned by Aristotle, modified and elaborated over the next centuries, was actually the transposition of the real and physical world onto the purely geometric and abstract model developed by Eudoxus of Cnidus in the first half of the fourth century BC Eudoxus' spheres, unlike Aristotle's, were not real physical entities but pure invention or mathematical artifice capable of intellectually accounting for perceptible appearance, and so able to justify and explain planetary motion, to "save the phenomena" or justify appearances.

The contrast between astronomy as a construction of hypotheses and astronomy intended to describe actual phenomena would prove significant. The split between cosmology and physics on the one hand, and an astronomy of mathematics and calculation on the other, began in Alexandria – the center of philosophical and scientific learning in the ancient world – with the theories of the great astronomer, Ptolemy, who lived there in the second century AD. His *Syntaxis*, also known as the *Almagest*, remained the basis of astrological and astronomical knowledge for over a thousand years.

Aristotle's spheres were actual solid, crystalline entities. The eccentrics and epicycles that Ptolemy – who always began his explanation of planetary motion with the statement "Imagine a circle" – described did not physically exist. They were, according to Proclus (410–485 AD), merely the simplest way to describe planetary movement. Ptolemy presented astronomy as a subject for mathematicians rather than physicists. However, the complex description of the universe that remained fairly fixed until the time of Copernicus cannot be traced to either of the theories discussed here. It was really a combination of Aristotelian physics and Ptolemaic astronomy set in a cosmology influenced to a great degree by NeoPlatonic mysticism, astrological views, the theology of the Church Fathers and Scholastic philosophers. The universe described by Thomas of Aquinas (1225–74) is proof of this, as is Dante's (1265–1321) depiction in the *Divine Comedy*, where heavenly bodies are endowed with all sorts of angelic powers.

Although an oversimplification, what follows is a brief list of those cosmological assumptions that had to be challenged and rejected in order to create a new astronomy.

1   The distinction in principle between celestial and terrestrial physics that resulted from dividing the universe in two parts; one perfect and the other subject to change.
2   The consequent belief in the circular motion of the divine planets.
3   The assumption that the Earth was immobile and lay at the center of the universe, which was supported by a series of seemingly irrefutable arguments (terrestrial motion would hurl objects and animals into the air) and confirmed by Scripture.
4   The belief in a finite universe and a closed world linked to the doctrine of natural position.
5   The conviction, closely related to the distinction between natural and violent motion, that it was not necessary to explain the resting state of a body whereas all movement had to be explained as either depending on the form or nature of a body, or as the result of a moving force that produces or conserves it.
6   The widening divide between physics and the mathematical theories of astronomy.

Over the course of a hundred years (from about 1610 to 1710) these presumed truths were debated, critiqued, and refuted. What emerged after a long and at times torturous process was a new image of the physical universe which reached completion in the work of Isaac Newton; the great construction known – until the time of Einstein – as "classical physics". Each rejection naturally involved a radical overturning of mental images and interpretive categories and brought with it a new way of looking at nature and man's place in it.

Vile Mechanics

Bachelardian obstacles – those having to do with knowledge and ways of "looking at the world" – were not the only ones to challenge early modern science during its difficult formative years. Values and opinions relating to social structure and the organization of labor also posed challenges, as did the prevailing social view of the scholar insofar as that view dominated the organizations in which knowledge was generated and transmitted.

Central to the great Scientific Revolution of the seventeenth century was the interpenetration between the technical and the scientific. For better or worse, that interpenetration marked all of western civilization and took on a form in the seventeenth and eighteenth centuries (which then spread the world over) that it did not have in the ancient and medieval worlds. The Greek word *banausia* means mechanical arts or manual labor. In Plato's *Gorgias*, Callicles declares that machine builders are despicable, and to be called *bánausos* is an insult. Furthermore, no man would let one marry his own daughter. Aristotle excluded the "manual workers" from his citizenship roster and differentiated them from slaves only because they served the needs of many while slaves served the needs of only one. The difference between slaves and freemen gave way to the difference between the technical and the scientific; between forms of knowledge that addressed the practical and useful as opposed to those that addressed the contemplation of truth. Disdain for slaves, who were considered inferior by nature, extended itself to the work they did. The seven *liberal* arts of the trivium (grammar, rhetoric, logic) and the quadrivium (arithmetic, geometry, music, astronomy) were called liberal because they were the very arts practiced by free *(liberi)* men as opposed to servants or slaves who practiced the mechanical or manual arts. According to Aristotle and the Aristotelian tradition, knowledge that did not serve external ends was the only kind which got at the essence of man. And to engage in philosophy required material well-being; the necessities of life needed to be taken care of. Though the mechanical arts were necessary for philosophy and embodied the assumptions upon which it was based, they were considered inferior forms of learning, immersed in the material and sensible world and associated with the practical and the manual. In Stoic and Epicurean philosophy, as in the later thinking of Thomas of Aquinas, the ideal of the sage was more consistent with the image of someone who dedicated himself entirely to contemplation in the expectation (for Christian philosophers) of achieving the blessedness of God's contemplation.

In light of these considerations, it is highly significant that many fifteenth-century authors extolled the virtues of the active life. Giordano Bruno praised the hands, and many sixteenth-century texts on engineering or machine-building defended the manual arts, a refrain that was repeated by Bacon and Descartes as well.

In one of the most famous technical treatises of the Renaissance, *De re metallica* (1556), Georg Agricola (Bauer) passionately defends metallurgy against charges of being "unworthy and vile" compared to the liberal arts. Metallurgy was considered by many to be servile labor and "shameful and dishonest for the free man, or the honest and honorable gentleman." Agricola argued that "miners" had to be experts in locating the right terrain, finding deposits of minerals, and distinguishing between the various types of rocks, gems, and minerals. Philosophy, medicine, the art of measuring, architecture, design, and law were all required for his art. Technical labor could not be separated from scientific work. To his opponents who based their arguments on the contrast between freeman and slave, Agricola pointed out that slaves had once practiced agriculture, contributed to architecture, and had even been famous doctors (Agricola, 1563: 1–2).

Guidobaldo del Monte presents a similar argument in *Mechanicorum libri* published in Pesaro in 1577. In much of Italy "the term mechanic is considered by some to be an insult, and being called an engineer is resented by others." The term mechanic instead means "a very important man who knows how to execute marvelous projects with his hands and heart." Archimedes was primarily a mechanic. To be a mechanic or an engineer "is the office of a worthy and well-bred person, and mechanic is the Greek word for something made with great skill and includes all buildings, instruments, and devices . . . or a skillfully wrought genius which requires science, art, and practice" (Guidobaldo, 1581: *Ai lettori*).

To fully understand this "defense" of the cultural value of technology one must remember that Richelet's *Dictionnaire français*, published in 1680, still defined *mécanique* as: "mechanic, used in the arts, means the opposite of liberal and honorable: it has the connotation of being lowly and rude, and unworthy of an honest man." Callicles' notion was still alive in the seventeenth century: *vile mechanic* was the sort of insult that caused gentlemen to draw their swords.

The debate about the mechanical arts, which peaked between the mid-sixteenth and mid-eighteenth centuries, was associated with several important themes in European culture. Artistic and experimental production, as well as the writings of engineers and skilled workers, reflected new views on labor, on the function of technical knowledge, and the importance of processes that artificially alter and transform nature. An untraditional view of the *arts* gradually emerged in the realm of philosophical discourse: some of the processes that skilled workers and artisans employed to modify nature aided the making of discoveries in the real world, and were in fact valuable in demonstrating (as would be stated in explicit controversy with traditional philosophies) "nature in movement."

It is only in this context that the attitudes underlying Galileo's great astronomical observations take on their full meaning. In 1609 Galileo pointed his telescope at the heavens. What was revolutionary was Galileo's *faith* in an instrument born in the shops of mechanics and improved upon through trial alone; for though it was recognized as having some military applications, it was ignored – when not disparaged – by the official scientific world. The telescope was invented by Dutch craftsmen. Galileo *rebuilt* it, presented it in Venice in August of 1609, and then offered it as a gift to the government of the Doges. Galileo did not consider the telescope simply one of many unusual instruments designed to entertain the court or be immediately used by the military. When he pointed it at the sky it was in a methodical way, and his scientific approach transformed it into a scientific instrument. In order to trust what the telescope revealed it was essential to believe that it did not *deform* vision but enhance it. One had to believe instruments were a source of knowledge, and abandon the deeply-rooted anthropocentric view that only man's natural and unaided sight could build knowledge. *Introducing instruments into the scientific world* and conceiving of them as sources of truth was no easy task. *Seeing*, in today's science, almost always means *interpreting signs generated by instruments*. Behind what we today *see* in the heavens lay a single gesture of intellectual courage.

Defending mechanics against charges of unworthiness and refusing to equate culture with the liberal arts and servile labor with practical activities led to the rejection of the traditional image of science and the end of a basic distinction between knowing and doing.

2
Secrets

"Margaritae ad porcos"

In the Gospel according to Matthew (7: 6), Jesus says: "Do not give dogs what is holy, and do not throw your pearls before swine, lest they trample them under foot and turn to attack you." For centuries, scholars interpreted this to mean that what was precious was select, and that truths should be kept secret for their spread could be dangerous.

The idea that there was a secret knowledge of essential things – the spread of which would be disastrous – for centuries formed a dominant paradigm in European culture. Only the spread of this paradigm of secrecy, and its persistence and historical continuity, could explain the controversial force found in so many of the theories advanced by the so-called founders of the modern world: they unanimously rejected the idea upon which this secrecy was founded; namely, the difference between the learned few, or "true men," and the *promiscuum hominum genus*, or ignorant masses.

Hermetic Knowledge

The communication and spread of knowledge, and public discussion of theories (both common practices today) have not always been considered values. Rather, they have *become* values. From the beginning of European intellectual history, the importance of communicating ideas has always clashed with the belief that the mysteries of knowledge should be available only to an elite few.

The *Secreta secretorum* (which was attributed to Aristotle) was widely known in the Middle Ages. Written as a series of letters to his disciple,

Alexander the Great, the philosopher reveals the secrets of medicine, astrology, physiognomy, alchemy, and magic that were reserved for his closest followers. More than 500 manuscripts of what Lynn Thorndike has described as "the most popular book of the Middle Ages" have been identified in European libraries. Though this secret literature remained outside the medieval universities, it was widely read by the greatest figures of the new intellectual culture. At the end of the thirteenth century, Roger Bacon proposed a *scientia experimentalis* which Lynn Thorndike accurately describes as two-thirds Hermetic and not transmittable to the ignorant masses: "The wise [. . .] have either omitted these topics from their writing, or have veiled them in figurative language [. . .]. Hence according to the view of Aristotle in his book of secrets, and his master Socrates, the secrets of science are not written on the skins of goats and sheep so that they may be discovered by the multitude" (Eamon, 1990: 336).

The Hermetic view of the world and of history was strongly tied to the Gnostic and Avveroistic notion that there were two kinds of people: the simple, ignorant masses and an elect few capable of discerning truths concealed in words and symbols and so privy to the sacred mysteries. This view was clearly articulated in the *Corpus Hermeticum*, 14 treatises from the second century AD that were translated by Marsilio Ficino (1433–99) between 1463 and 1464. The texts had already circulated widely in manuscript form before their publication in sixteen editions between 1471 and the sixteenth century. Ficino attributed the treatises to Hermes Trismegistus (an attribution which persisted throughout the sixteenth and early seventeenth centuries), a legendary Egyptian god who lived at the time of Moses and was the indirect teacher of Pythagoras and Plato. The treatises rekindled an interest in magic in the late fifteenth and sixteenth centuries, and continued to influence European culture through the middle of the seventeenth century. The *Corpus* placed the great legacy of ancient and medieval magical and astrological knowledge into a broad, organic Platonic–Hermetic scheme. A search for the Unity that underlies differences characterizes the text; the desire to reconcile differences; the need for complete pacification in the One-All.

To the men of that day, there was a fine and fleeting line between natural philosophy and mystic knowledge; between a man who knows and experiments with nature and a man like Faust, who sells his soul to the devil for knowledge of the natural world and power over it. *Nature*, in the magical tradition, was not just a continuous and homogeneous matter that filled space; it was animated by a soul, a source of internal and spontaneous activity. That soul-substance was, for the Ionic thinkers of the fifth century BC, "full of gods and monsters." Every object overflowed with occult sympathies that connected it to the All. Matter was impregnated with the divine. The stars were living divine creatures. The world was the image or reflection of God and man was the image or the reflection of the world. Things in the greater world or *macrocosm* corresponded precisely to things in the miniature world or *microcosm* (of which man himself was a prime example and reference). The plants and forests were the hair and fur of the world, the rocks its bones, and subterranean waters its veins and blood. And man, at the center of the world, represented its heart. Insofar as he was the reflection of the universe, man was capable of revealing and grasping those secret relationships. Magicians and sorcerers were figures able to penetrate the infinitely complex reality of a system of similarities and Chinese nesting boxes that led to the All and enclosed it. The sorcerer was privy to the chain of connections that flowed from the top down, and – through invocation, numbers, images, names, sounds, chords, and talismans – knew how to construct an unbroken chain of ascending links. Love was the *nodus* or *copula* that firmly bound one part of the world to the other, and which Ficino saw as "connected one to the other by a sort of reciprocal charity, [. . .] limbs of a single creature, united one to the other by the communion of a single nature." Vitalism, animism, organicism, and anthropomorphism were components of Hermetic thought, the focus of which – as Freud and Cassirer clearly saw – was the idea of an identification between the self and the world and the "omnipotence of thought."

The magical sphere was compact and totalitarian; neither easily undermined nor open to repudiation. And the wonderful deeds performed by the sorcerer confirmed that he belonged to the rank of the elect. The distinction between the elect and the masses was a necessity of secret knowledge, in which truths must remain concealed to the point of seeming unknowable. For its methods were so difficult that most men were incapable of understanding them, and its terminology was necessarily ambiguous and allusive because the procedures were so complicated, and born of the need to restrict that knowledge to the few. The fact that comprehension of the truth was achieved despite language rather than through language further confirmed that truth belonged to a select few.

It has often been observed that magic resembles both psychology and religion; yet it is neither psychology nor religion nor for that matter mysticism. Just as astrology embodied sophisticated calculations and anthropomorphic vitalism, magic and alchemy embodied mysticism as well as experimentalism. Renaissance books about magic today appear to be a strange combination of subjects. A single manual includes pages on optics, mechanics and chemistry, formulas for medicinal cures, technical instructions on how to build machines and mechanical games, secret codes, recipes for food as well as rat poison, advice to fishermen, hunters, housewives, and prestidigitators, information on hygiene, aphrodisiacs and sex, and references to metaphysics, mystical theology, traditional Egyptian wisdom, the Bible and classical and medieval philosophy. Moreover, magic – as associated with Giordano Bruno, Cornelius Agrippa, and Tommaso Campanella – was deeply tied to ideas of cultural reform, millenarianism, and hopes for radical political renewal.

The language of alchemy and magic was ambiguous and allusive precisely because the idea that secret knowledge could ever be expressed clearly and simply was incomprehensible. Language was structurally and deliberately full of semantic twists and turns, metaphor, analogy, and allusion. For instance, the alchemist Bono of Ferrara wrote that "no ancient could ever achieve the divine subject of this art through his natural intellect, nor according to natural reason alone, nor according to experience because it – like a divine mystery – is above reason and experience" (Bono of Ferrara, 1602: 123).

The alchemist did not write about gold and sulfur in a concrete state. An object was never simply itself; it was also the symbol of something else, receptacle of a transcendent reality. For this reason, a chemist who today reads a paper on alchemy "has something of the same experience as a mason hoping to learn something practical from a work of freemasonry" (Taylor, 1949: 110). By virtue of their understanding the secrets of Art, initiates "prove that they belong to the group of the enlightened." Lovers of Art, wrote Bono of Ferrara, "understand one another as if speaking a common language that is incomprehensible to others and known only to themselves" (Bono of Ferrara, 1602: 132). In *Magia adamica*, Thomas Vaughan stated that knowledge is made up of visions and revelations, and only through divine enlightenment can man reach a complete understanding of the universe (Vaughan, 1888: 103).

The distinction between *homo animalis* and *homo spiritualis*, the separation of the simple from the wise, became the identification of the ends of knowledge with salvation and individual perfection. Science corresponded to the purification of the soul and became a means for escaping one's destiny on earth. Intuitive knowledge was superior to rational knowledge; an occult knowledge of things was equated with freedom from evil: "I have written this work for you, sons of knowledge. Read it carefully and reap the wisdom that has been scattered throughout. What may have been concealed in one part has been made manifest in another [. . .]. I have written this for you alone of pure spirit and chaste mind, whose uncorrupted faith fears and honors God [. . .]. You alone will find that which I have meant for you alone to find. The secrets cloaked in mystery cannot be revealed without occult intelligence, so all of magical science will course through you as will the virtues of Hermes, Zoroaster, Apollonius, and the other practitioners of wondrous things" (Agrippa, 1550: I, 498).

Ad laudem et gloriam altissimi et omnipotentis Dei, cuius est revelare suis praedestinatis secreta scientarum (For the promise and glory of almighty God, who reveals to the elect the secrets of knowledge): The theme of secrecy occurs in the opening pages of the *Picatrix* and continues throughout. Philosophers skillfully hid magic behind secret words, and this they did for altruistic motives: *si haec scientia hominibus esset discoperta, confunderent universum* (If this knowledge were revealed to all men, it would confound the universe). Science had two sides; one open and the other closed. The obscure science was profound, for the very words that described the natural order were those given to Adam by God, and comprehensible to only a select few (Perrone Compagni, 1975: 298).

What is striking about the idea of secrecy is not the formulas themselves, but their immutability. The same authors, citations, and examples recur in occult texts through different periods in history. Cornelius Agrippa, for example, tells us that Plato forbade the disclosure of the mysteries, Pythagoras and Porphyry bound their disciples to secrecy, Orpheus as well as Tertullian demanded vows of silence, and Theodotus was blinded because he tried to penetrate the mysteries of Hebrew scripture. Indians, Ethiopians, Persians, and Egyptians spoke through riddles. Plotinus, Origen, and Ammonius' other disciples vowed never to reveal their teacher's dogma. Christ himself obscured his words in such a way that only his most trusted disciples could understand them, and he explicitly prohibited giving consecrated meat to dogs and pearls to swine. "Every magical experience abhors the public and wants to remain hidden; it is fortified by silence and destroyed when declared" (Agrippa, 1550: I, 498).

Truth was transmitted personally by "the whispers of tradition and oral discourse." Direct communication between teacher and disciple was privileged: "Without a trusted and expert teacher, I do not know if it is possible to divine understanding simply by the reading of a text [. . .]. These things are not entrusted to letters or written with the pen, but infused from spirit to spirit through sacred words" (ibid.: II, 904).

Public Knowledge

For a thousand years (or the ten centuries of the Middle Ages) the dominant figures in western culture were the saint, the priest, the doctor, the university professor, the soldier, the craftsman, and the sorcerer. Later, the humanist and courtier took their places alongside these figures, and then between the mid-sixteenth and mid-seventeenth centuries new ones were added: the *mechanic*, the *natural philosopher*, and the *virtuoso*, or free experimenter, individuals who did not seek after sanctity or literary immortality, nor the production of miracles to enthrall the masses. Scientific knowledge developed in a climate of bitter controversy with monastic, scholastic, humanistic, and scholarly learning. John Hall, in a 1649 motion to Parliament, charged that universities did not teach chemistry, anatomy, languages or experiments, and students were like mummies who had been taught science 3,000 years ago in hieroglyphics awakening from a long sleep. The occult knowledge of sorcerers and alchemists had been fiercely contested by mechanics and engineers even before philosophers opposed it. In the *Pirotechnia* (1540), Vanoccio Biringuccio accused alchemists of not being able to *codify their methods* and concentrating only on ends, offering "authoritative testimony rather than possible reasons of demonstrable effects. Some cite Hermes, Brother Arnoldo, or Raymond, others Geber, Ockham, Craterius, or the saintly Thomas Aquinas, and even a certain Franciscan brother Elias; these authors, invoking the dignity or even sacredness of their philosophical knowledge, expect a sort of respectful faith of their readers or that their listeners either remain silent like ignoramuses or else confirm whatever they say" (Biringuccio, 1558: 5r). Unlike Biringuccio, Georg Agricola (Georg Bauer) was very well read. *De re metallica* (1556) – a text that was kept chained to the altars of New World churches as a universal manual – forcefully addressed questions of secret knowledge that are, in principle, impossible to decode "because the writers upon these things use strange names, which do not properly belong to the metals, and because some of them employ now one name, and now another, invented by themselves, though the thing itself changes not" (Agricola, 1563: 4–5).

Later on, a series of social and economic events reinforced the value of "secrecy" in the world of mechanics as well. Many Renaissance artisans and engineers insisted upon the right to keep their own inventions secret, but their motivations were purely financial and unrelated to any idea that the masses were unworthy. The first patents date back to the early fifteenth century; and their use increased dramatically in the sixteenth century (see Eamon, 1990; Maldonado, 1991).

During the tumult of the religious wars in Europe, the men who first called themselves "natural philosophers" constructed smaller and more tolerant societies within the greater societies in which they lived. "When I lived in London," wrote John Wallis in 1645, "I had the opportunity to meet some people who were working on what we today call new or experimental philosophy. We excluded theology from our discourse, and were more interested in physics, anatomy, geometry, statics, magnetism, chemistry, mechanics, and natural experiments."

The members of the first Academies wanted to protect themselves primarily from two things: politics and the intrusiveness of theology and the

Church. The Accademia dei Lincei "has in particular banned from their subjects of study any argument outside the natural and mathematical, and eliminated political issues." The Royal Society requests "a close, naked, natural way of speaking and clear senses [. . .]: and preferring the language of artisans, countrymen, and merchants before that of wits or scholars" (Sprat, 1667: 113).

With regard to scientific academies and societies, several points should be emphasized. First, the meetings of learned men were governed by a code of conduct, and it was a point of principle to exercise a critical attitude towards all statements. Truth was not bound to the authority of the person who declared it, but only to the proof of the experiment and force of the demonstration.

Secondly, it should be remembered that all adepts of the new science favored linguistic rigor and non-allusive terminology. This position went hand-in-hand with the refusal to make any basic distinction between scholars and laymen. Theories had to be fully communicable and experiments continually repeatable. According to William Gilbert, "sometimes we use new words. Though not to obscure like the alchemists do, but so that hidden things can be fully understood" (Gilbert, 1958: Praefatio). In the celebrated opening of *Discourse on method*, Descartes claimed that good sense "is the best thing the world has given us." The ability to judge what is good and distinguish what is true from what is false (which constitutes reason) "is by nature equal in all men." In addition, reason, which makes us different from animals, "is whole in each of us." The method pursued by Hobbes that leads to science and truth is made for all men: "If you would like," he says to the reader in the preface to *De corpore*, "you too can use it." Bacon also believed that scientific method tended to diminish the differences between men and make their intelligence equal.

Ritual magic, wrote Bacon, goes against the divine commandment that men earn their bread by the sweat of their brow, and "proposes that by following a few simple and easy steps man can achieve the noble ends that God demands he acquire only by his own labor." Inventions, he continues, "are cultivated by few and in absolute and almost religious silence." All who criticized and opposed magic emphasized the "priestly" nature of magical knowledge and the commingling of science and religion that characterized the Hermetic tradition.

Father Mersenne asked himself why followers of alchemy were not willing to study the results of their discoveries "without mystery and secrets?" (Mersenne, 1625: 105). Francis Bacon's high opinion of the intellectual courage shown by Galileo in his astronomical discoveries included praise of his intellectual honesty: "men of this sort have continued to account for every single point of their research in a honest and perspicuous manner" (Bacon, 1887–92: III, 736). Those who get lost following unusual paths, wrote Descartes, are less readily excused than those who err in the company of others. In this "darkness of life," believed Leibniz, it was necessary to walk together because scientific method was more important than individual genius and the goal of philosophy was not to improve the intellect of the individual but that of all men. Leibniz, Hartlib and Comenius, each in his own way, referred to the ideal of the "advancement of learning" or the development and spread of knowledge. To the author of the *Pansophiae prodromus*, "the people's desire for schools" was characteristic of this new age. He believed that this desire produced "the great increase of books in every language and every nation so that even children and women could become familiar with them [. . .]. Now finally there emerges the steady drive of some people to perfect the method of learning to such a degree that anything worth learning can be easily instilled into minds. If this effort (as I hope) is successful, we shall find the sought-for path of rapidly teaching everything to everybody" (Comenius, 1974: 491).

It was inevitable that over the course of the seventeenth century the battle in favor of a universal knowledge that could be comprehended by everyone because it could be communicated and constructed by everyone shifted from the level of the ideas and projects of the intellectual to those of the institution. Sprat had this to say about the membership of the Royal Society: "As for what belongs to the Members themselves, that are to constitute the Society: it is to be noted that they have freely admitted men of different religions, countries, and professions of life. [. . .] For they openly profess not to lay the foundation of an English, Scotch, Irish, Popish, or Protestant philosophy; but a philosophy of mankind." About the design of the organization he wrote: "They have tried to put it into a condition of perpetual increasing; by settling an inviolable correspondence between the hand and the brain. They have studied to make it not only an Enterprise of one season, or of some lucky opportunity, but a business of time; a steady, lasting, popular, and uninterrupted work. They have attempted to free it from the artifice, and humors, and passions of sects; to render it an instrument, whereby mankind may obtain a dominion over *things* and not only over one another's *judgments*. And lastly, they have begun to establish these reformations in philosophy, not so much by any solemnity of laws, or ostentation of ceremonies, as by solid practice and examples: not by a glorious pomp of words; but by the silent, effectual, and unanswerable arguments of real productions" (Sprat, 1667: 63, 62).

The Hermetic Tradition and the Scientific Revolution

Scholarship in the last half century has increasingly revealed that the Hermetic–magical tradition had an important influence on many of the leading exponents of the Scientific Revolution. In early modern Europe, magic and science formed a web not so easily unraveled. Today, the Enlightenment-era positivist image of scientific knowledge marching triumphantly through the darkness and superstition of magic has definitively been laid to rest.

In defending heliocentricity, Copernicus appealed to the authority of Hermes Trismegistus. William Gilbert invoked Hermes and Zoroaster when he linked his theory of earthly magnetism to that of a totally animate universe. Francis Bacon's doctrine of *forms* was strongly affected by the language and the models used by alchemists. Kepler was well acquainted with the *Corpus Hermeticum*, and both his belief in a mystic harmony between geometrical structures and the universe and his theory of the celestial music of the planets were deeply imbued with Pythagorean mysticism. Tycho Brahe saw astrology as a legitimate application of his scientific work. Descartes, whose philosophy was considered the symbol of rational clarity by moderns, as a young man preferred the products of the imagination to those of reason. Like many sixteenth-century sorcerers he took pleasure in constructing self-moving mechanisms and "shadow gardens"; like many exponents of magical Lullism, he insisted on the unity and harmony of the cosmos. These are themes that also recur, albeit differently, in the work of Leibniz, whose ideas derive in part from the tradition of Hermetic and cabalistic Lullism. In fact, the Leibnizian notion of harmony was based on a passionate reading of texts that could hardly be thought of as "scientific." William Harvey's *De motu cordis*, which praises the heart as the "sun of our microcosm," reflects Hermetic and sun worship themes of the fifteenth and sixteenth centuries. There is a precise relationship between Harvey's definition of *ovum* or egg (neither fully alive nor entirely void of vitality) and Ficino's (as well as other alchemists and Paracelsians) definition of *astral body*. Even the Newtonian concept of space as a *sensorium Dei* revealed the influence of Neoplatonic currents and the Jewish Cabala. Newton not only read and summarized texts on alchemy, but he devoted many hours of his life to research that was alchemistic in nature. His written work also revealed his faith in a *prisca theologia* (the central theme of Hermeticism), the truth of which could be "proven" through new experimental science.

What caused "sorcerers" and "scientists" to finally part company in the late sixteenth and early seventeenth century is more complex than a simple appeal to experience or revolt against *auctoritates*. Gerolamo Cardano, for example, was a successful mathematician and Giambattista della Porta an important contributor to the history of optics. Certainly the calculations made by many astrologers were much less questionable than the mathematical digressions of Hobbes and Paracelsus, and less "scholastic" than those of Descartes.

For Bacon, to humbly leaf through the great book of nature meant to renounce the building of entire systems of natural philosophy on too-fragile conceptual and experimental bases. Francesco Patrizi and Peter Sørensen (Severinus), Bernardino Telesio, Giordano Bruno, Tommaso Campanella, and William Gilbert were, in his opinion, a series of philosophers who arbitrarily defined the subjects of their worlds. On the other hand, he believed the Veronese physician Girolamo Fracastoro (1483–1553) to have been a man of honest and free judgment. In *De symphathia et antipathia rerum* (1546), Fracastoro addressed a series of common problems (e.g., why the magnet points north, how the remora or sucking fish stops a ship, etc.), but underlying his study of contagions was an inquiry into "sympathy and antipathy." Fracastoro observed that contagions had till then been interpreted as manifestations of occult properties. Rather than investigate the principles of contagion, how it occurs, the differences in severity of contagious diseases, and the difference between a contagious diseases and poisoning, mysterious causes had generally been invoked. Philosophers had dedicated themselves to explaining "universal causes" while neglecting the study of "specific and particular causes" (Fracastoro, 1574: 57–76). "Sympathy" was better explained as a *force* rather than in terms of the mysterious *nature* of bodies, and this theoretical shift made Aristotelian theory useless. Citing Democritus, Epicurus, and Lucretius, Fracastoro accepted the idea that the *effluxiones* of bodies hold the principle of attraction. The attraction between two bodies was the result of the reciprocal transmission of minute particles from body A to body B. The sum of these particles formed a whole but one that varied across space: the particles next to the two bodies and those located between them do not have the same density and rarefaction. Therefore, movements that tend to achieve equilibrium or the maximum consent of the parts with the whole are produced in the "cloud of atoms." These adjusting movements cause the two bodies to move toward each other, and in some cases, unite.

In the sixth chapter of *De contagionibus et contagiosis morbis* (1546), Fracastoro declared that "contagion that occurs at a distance cannot be attributed to occult properties" (Fracastoro, 1574: 77–110). Some contagion occurs through simple contact (scabies and leprosy); others by means of a vehicle, such as bedsheets or clothing; yet others are spread from a distance by invisible "seeds." Fracastoro – also known for his famous poem in Latin *Syphilis sive de morbo gallico* (1553) – showed other signs of a break with the occult in his pamphlet entitled *De causis criticorum diebus*. The critical days or "crises" for illnesses definitely occur on particular days, though they are impossible to establish by means of strict numerical relationships (as the "Pythagorean philosophers" suggest) or on the basis of a cause–effect relationship with the motion of the planets (as astrologers do). Doctors have been mistaken in not conducting painstaking experimental research into these questions, and in "letting themselves be seduced by the ideas of astrologers" (ibid.: 48–56).

In the more general philosophical context, there were however differing views on the solidarity between things, and on sympathy and antipathy. The notions could be *applied* differently, linking them to either a mystical vision of reality or using them as criteria or hypotheses for an "experimental" inquiry into nature.

Secrets and Public Knowledge

To fully understand the seemingly obvious difference between Renaissance *magic* and modern *science*, one must consider not only content and method, but also the image of knowledge and of the *scholar*. We certainly still live in a world full of secrets, and there are many who study and practice the *imperii arcana*. There are also many charlatans, and they too have their place in the history of science. Nevertheless, it should be pointed out that after the first Scientific Revolution, there was not, nor could there have been, *praise* for or a positive view of dissimulation in the scientific literature or literature about science (an observation which, for example, still does not apply to the world of politics). To dissimulate, or not make public one's own opinions, simply implies trickery or betrayal. Scientists working as a community may indeed pledge secrecy, but the pledge is usually imposed upon them. And when such restrictions are imposed, scientists inevitably protest against them or, as has occurred in more recent times, rebel outright. The fact that "Kepler's laws" are called "Kepler's" has nothing to do with possession, and simply serves to perpetuate the memory of a great figure. For science itself, and within the scientific world, secrecy became a liability.

3

Engineers

Practice and Discourse

In the preface to *Discours admirables* (Paris, 1580), Bernard Palissy rails against the professors of the Sorbonne and asks if it is possible for a man to be knowledgeable about natural effects without having read books written in Latin. Palissy was an apprentice glassmaker who reached great celebrity and ended in ruin while trying to discover a secret white pottery enamel. Over the course of his adventurous life he designed a number of machines which went unbuilt and more than once risked death by starvation and execution. Palissy died in the Bastille in 1589 or 1590. To the question he posed his readers, Palissy answered by stating that yes, practice and experience could show that the teachings of philosophers (even the most well-known) could be false. His workshop and museum of natural and artificial objects could teach more philosophy than any Sorbonne course in ancient philosophy (Palissy, 1880).

A year after Palissy's *Discours* were published, a pamphlet titled *The New Attractive, Containing a Short Discourse of the Magnet or Lodestone* was published in London. This text about magnetism and the dip of the magnetic needle – later to be of great use to William Gilbert – was written by Robert Norman (*fl.* ca. 1560–96), an English sailor who after twenty years at sea had turned to the manufacture and sale of compasses. Norman called himself an "unlearned mathematician" who had accumulated a great deal of data in the course of his maritime career. He decided then to risk his reputation and bear the insults of his enemies so that he could propose for public consideration the fruits of his labor. Norman claimed to work for the glory of God and for England. The reader was reminded that the author was but a simple mariner, and unable to debate with logicians or give a satisfactory explanation for the causes of earthly magnetism. He had a clear sense of the difference and the fundamental contrast between his research and that of "lettered men." These latter developed refined concepts and expected all mechanics to hand over their discoveries to them. Fortunately, concluded Norman, "there are in this land diverse Mechanitians, that in their severall faculties and professions have the use of those Arts at their finger endes. And can applye them to their severall purposes, as effectionately and more readily than those who woulde most condemne" (Norman, 1581: Preface).

Ideas of this sort quickly penetrated the scholarly world as well. For example, Juan Luis Vives (1492–1540), friend of Erasmus and of Thomas More, a tutor at the English court and a learned man who addressed his writings to a refined public of humanists, expressed the same idea less naively but just as energetically. In *De tradendis disciplinis* (1531), Vives invited European scholars to seriously consider the technical problems of machines, weaving, agriculture, and navigation. He urged them to overcome their traditional disdain for such things, and enter workshops and factories, ask craftsmen questions and try to understand the details of their work. The science of nature, Vives declared in *De causis corruptarum artium* (1531), was not the exclusive domain of philosophers and dialecticians, but better known by mechanics who had never developed imaginary constructs such as *forms* and *thisness*

Palissy, Norman, and Vives – though at different levels and with different intentions – voiced the same demand for a knowledge that favored observation and empirical research over an exclusively verbal one. This same theme runs through one of the greatest texts of the new science, *De corporis humani fabrica* (1543) by Andreas Vesalius. Vesalius enthusiastically argued against the dichotomy that had developed in the medical profession, in which the professor lectures from a book at a careful distance from the corpse while the dissectionist, who is ignorant of any theory, is reduced to the status of a butcher.

The texts I have referred to above go back to the half century between 1530 and 1580. Common themes are sounded in the writings of a Parisian artisan, an English sailor, a Spanish philosopher, and a Flemish scientist linked to the Italian cultural tradition: that the methods used by artisans, artists and engineers are valuable to the progress of knowledge, and that such methods should be recognized as possessing their own cultural dignity (Rossi, 1970: 1–11).

Engineers and Machines

Many of the Renaissance translations of classical texts were aimed expressly at a growing public of artisans. Jean Martin translated Vitruvius's architectural treatises (first century BC) in 1547 into French for workers and other people who could not read Latin. Walter Rivius translated the same text into German in 1548, and addressed an audience of artisans, craftsmen, stonecutters, architects, and weavers. The many commentaries on Vitruvius, including a notable 1556 translation by the Venetian aristocrat Daniele Barbaro, offer a clear example of the meaning and importance of such "representations" of the classical texts.

Many highly-skilled artisans came into contact with ancient learning, finding the answers to their questions in the work of Euclid, Archimedes, Hero, and Vitruvius. As we know, fifteenth- and sixteenth-century literature was extraordinarily rich in technical treatises; while some of these were real manuals, others were random collections of observations and reflections on the work of artists or "mechanics" or the processes used in the various arts. This vast body of literature includes the works of engineers, artists, and master craftsmen: Filippo Brunelleschi (1377–1446), Lorenzo Ghiberti (1378–1455), Piero della Francesca (1406–ca. 1492), Leonardo da Vinci (1452–1519), and Paolo Lomazzo (1538–1600); the treatises on instruments of warfare by Konrad Keyser (1366–1405); architectural texts by Leon Battista Alberti (1404–72), Francesco Averlino "Il Filarete" (1416–70), and Francesco di Giorgio Martini (1439–1502); Valturio of Rimini's book on military machines (published in 1472 and reprinted in Verona in 1482 and 1483, in Bologna in 1483, in Venice in 1493 and four times in Paris between 1532 and 1555); Albrecht Dürer's (1471–1528) two treatises on descriptive geometry (1525) and fortifications (1527); *Pirotechnia* by Vanoccio Biringuccio (ca. 1480–ca. 1539), first published in 1540 and twice reprinted in Latin, three times in French, and four times in Italian; a work on ballistics in 1537 by Niccolò Fontana a.k.a Tartaglia (ca. 1500–57); two treatises on mining engineering by Georg Agricola (1494–1555) published in 1546 and 1556; the *Théâtres des instruments mathématiques et méchaniques* (1569) by Jacques Besson; *Diverse et artificiose machine* (1588) by Agostino Ramelli (1531–90); *Mechanicorum libri* (1577) by Guidobaldo del Monte; three books on mechanics by Simon Stevin or Stevinus (1548–1620); *Machinae novae* (1595) by Fausto Veranzio (1551–1617); *Novo teatro di machine et edificii* (1607) by Vittorio Zonca (1568–1602); and the navigational treatises of Thomas Hariot (1560–1621) and Robert Hues (1553–1632), published respectively in 1594 and 1599.

Universities and convents were no longer the only producers of culture. A new sort of knowledge was born that had to do with the design of machines, the building of offensive and defensive weaponry, the construction of fortresses, canals, dikes, and the extraction of metals from mines. The people behind this type of knowledge – engineers or artist/engineers – acquired prestige equal to if not greater than that of physicians, magicians, court astronomers, and university professors. Leon Battista Alberti – painter, sculptor, architect, urban planner, and sophisticated humanist – believed that mathematics (the theories of proportions and perspective) was the common ground of the painter and the scientist. Painting is a science, and the perspective view used by painters is a science. He also felt the work of the architect united "reason" and "rule" with "action", and any praise for the architect immediately became a paean to engineering techniques that enable man to drill holes into mountainsides, displace enormous masses of water and rock, drain swamplands, regulate the course of rivers, and build ships, bridges and instruments of war.

The Workshop

As Antal reminds us, in the fourteenth century art was still considered a manual skill (Antal, 1947). Almost all early fifteenth-century artists had humble origins and were from families of peasants, artisans, and the lower middle class. Andrea del Castagno was the son of a peasant, Paolo Uccello of a barber, Filippo Lippi of a butcher, and Pollaiuolo, as his name suggests, was the son of a poultry vendor. At the beginning of the century, Florentine sculptors and architects belonged to a minor guild of masons and carpenters while painters were members of the major guild of doctors and chemists, together with the lower-ranked house painters and color grinders. Workshops produced everything from masterpieces to coats of arms, banners, inlaid work, tapestry and embroidery patterns, and objects in both terracotta and gold, and an apprenticeship began with manual jobs such as grinding colors and preparing canvases. In addition to designing and constructing buildings, the architect also designed mechanical instruments and weaponry, as well as the complicated "machinery" and mechanisms used in processions and special events.

By the middle of the sixteenth century – the age of Vasari – the lower sorts of artisanal tasks no longer seemed compatible with artistic dignity. The legendary moment when Charles V knelt to pick up a paintbrush dropped by Titian symbolizes the artist's change in status. However, well before artists were known as "geniuses" or authors of immortal masterworks, the fifteenth-century Florentine workshop produced the fusion of manual labor and theory. Some workshops (for example,

Lorenzo Ghiberti's during the manufacture of the doors of the Florentine Baptistry) became virtual industrial laboratories and turned out painters, sculptors, engineers, technicians, and designers and builders of machinery. Apprentices simultaneously learned to mix colors, cut stones, cast bronze, paint and sculpt, while studying some anatomy, optics, geometry, and perspective. The education of an "unlettered man" was a practical education based on a variety of sources sprinkled with bits and pieces of classical science quoting Euclid and Archimedes. The empirical knowledge of someone like Leonardo da Vinci was the product of just such an environment.

Leonardo da Vinci

Leonardo da Vinci (1452–1519), painter, engineer, builder and designer of machines, "unlettered" man and philosopher, became the modern symbol of the multi-talented man, of someone who bridged the traditional divide between the mechanical and liberal arts, practice and theory, manual and intellectual labor. His interests as a young man derived from the practices of the fifteenth-century workshop, and his working knowledge of materials led him to insist that practice and theory be united. Science which "begins and ends in the mind" has no truth because purely mental discourse excludes "experience, without which there is no certainty," and it is equally true that there is no certainty without the application of mathematics. One who falls in love with practice without science "is like a helmsman entering a canal without a tiller or a compass, uncertain of where he is going" (Solmi, 1889: 84, 86). And it is senseless to criticize Leonardo for ambiguity or uncertainty. Defending the union of practice to theory meant sometimes criticizing the supporters of pure theory and at other times taking to task the opponent who, according to Leonardo, "does not want much science because practice is enough for him." Leonardo da Vinci joined the painter's guild in 1472 and remained in Verrochio's workshop until 1476.

In 1478, Leonardo was called to Milan by Ludovico Sforza to work as a sculptor and founder. After preparing a report on the military defense of Tuscany for the Count of Ligny, he was forced to flee Milan in 1499 after the fall of the Sforzas. He took refuge in Mantua, and that same year was hired by the city of Venice as a military engineer. After an "itinerant" period (including a sojourn in Florence), he was hired by Cesare Borgia in 1502 as a military engineer. As he traveled throughout central Italy, Leonardo kept a notebook (known as Manuscript L) in which he recorded and sketched everything that was of interest to him. When Borgia fell, Leonardo returned to Florence in 1503 and painted the *Mona*

Lisa and the unfinished *Battle of Anghiari*. The great project to deviate the course of the Arno River and build a port in Florence was interrupted by war between Florence and Pisa. In 1506, Leonardo was back in Milan in the services of the king of France, and he organized the celebratory festivities to mark the arrival of Louis XII. He remained in Milan until 1513, when the French retreated, and then moved to Rome to serve in the papal court of Leo X. In 1516, Leonardo joined the French court of Francis I as an engineer, architect and mechanic, and remained there until his death, without ever having returned to Italy.

Leonardo grew increasingly theoretical as he grew older, especially during his second Milanese period (Brizio, 1954: 278). Yet while it is true that he made his complex hydraulic designs in this period, it is certainly not possible to seek in the thought of this artist the foundation of experimental methods and a new science. Not unjustly, after so much insistence on the "miracle" of Leonardo, commentators have recalled his utter disdain for typography and printing, and emphasized the fact that the evaluation that was made of Leonardo's codices at the time of their publication depended on the scarce or nonexistent data concerning the actual state of sixteenth-century scientific knowledge. Leonardo's research, though rich in dazzling intuitions and brilliant insights, never went beyond the level of experiments made for the sake of *curiosity*, and so did not become systematic, a key characteristic of modern science. Leonardo's investigations, wavering between experiments and notation, appear fragmented and pulverized into a series of brief notes and random observations that he kept for himself in a personal and obscure notation that was deliberately nontransmittable. Leonardo, who was always intrigued by a particular problem, had no interest in working on a systematic body of knowledge. He also was unconcerned with communicating, explaining and proving his discoveries to others, other fundamental qualities of what we call science. From this point of view, even the many famous machines he designed re-acquire their real dimensions. Rather than being designed to alleviate the labor of men and increase their domination over the world, they appear to have been designed as tools of escapism: festivals, entertainment, mechanical surprises, etc. It was no accident that Leonardo was more involved with the *elaboration* rather than the *execution* of his projects. His machines risked becoming "toys" while the concept of "force" (which was insisted upon) was certainly linked more to the Hermetic and Ficinian theme of universal animation than to the nature of rational mechanics.

Nonetheless, we do need to remember that throughout Leonardo's fragments we constantly encounter ideas that will return with great force, in different contexts, within the culture of the modern era: the idea of a necessary link between mathematics and experience and the important advance that conception represented; an unwavering opposition to the vain pretexts of alchemy; the invective against "the reciters and devotees of the works of others"; objection to the appeal to authority, the practice of those who rely on memory rather than their ingenuity; the image of nature which "does not break its own rules," but obeys a wonderful and inexorable chain of causes; the conviction that experience is capable of "silencing warring tongues" and "the eternal uproar" of the Sophists. And one can trace precise links to Galileo's "certainty produced by the eyes" and "doctors of memory," his description of nature as "deaf to our vain wishes" producing instead its effects "in a manner that is unfathomable to us." In the same vein, Bacon's rejection of purely empirical knowledge, his image of man as the master of nature only if he is able to obey its inexorable laws

The long-dominant image of a sort of "childhood of science," of which Leonardo da Vinci is the prime representative, is undoubtedly to be rejected. But the image of Leonardo as an admirable "precursor" and the "Leonardo miracle" also need to be explained. The metaphor of childhood – different from the notion of "precursor" – remains a rich one. The great choices that underlie modern science (mathematization, corpuscularism, mechanization) led, using today's terminology, art and science to follow different paths, to move along strongly divergent trajectories, and to progressively become distant from one another. It makes no sense any longer to attempt to bring them closer together or weld them once again into one. Leonardo's drawings and paintings are not however the simple tools of a scientific study that finds its methodology elsewhere. Many of his drawings of rocks, plants, animals, clouds, body parts, faces, movement of air and water are themselves "acts of scientific knowledge, or critical inquiries into the natural world" (Luporini, 1953: 47). What remains of Leonardo's work – his notes, drawings, and the unique and extraordinary fusion of text and drawing – allows us to look upon another world: on those men and those times in which a closeness, a copenetration (impossible and illusory today), between science and art seemed possible and indeed was at times achieved.

"Discourse" and "Construction"

Biringuccio's *Pirotechnia* (1540) is one of the greatest technical treatises of the sixteenth century. In order to remain true to a descriptive ideal, Biringuccio refrained from all rhetorical embellishment. He believed that the alchemist belonged to a group of people who hid their considerable ignorance of the subject matter they dealt with behind "a thousand fables." They have no capacity to develop "methods," too great a desire to become rich, and in looking too far in the distance fail to see that which is "intermediate" (Biringuccio, 1558: 6v, 7v). Unlike Biringuccio, Georg Agricola was a man of vast culture and broad interests. Born in Glachau (Saxony) in 1494, he studied in Leipzig, Bologna, and Venice. In 1527 he began to practice medicine at Joachimstal in Bohemia, at the time the most important mining region in Europe. Bauer was the burgermeister of Chemnitz and carried out several important diplomatic missions for Emperor Charles and King Ferdinand of Austria; he was respected by Erasmus and Melanchthon. *De ortu et causis subterraneorum* and *De natura fossilium* were among the first systematic treatments of geology and mineralogy. Published in 1556, a year after the author's death, Agricola's *De re metallica* remained for two centuries the definitive work on mining technology. In Potosì, which supplied gold and silver to all of Europe, Agricola's treatise was considered a sort of Bible and kept on church altars so that miners could perform their devotions every time they came to solve a technical problem. The twelve books that make up the treatise deal with every aspect of the processes of extraction, fusion and the treatment of metals, and discuss the location and direction of veins, the machinery and tools of the trade and how to use them, how to test gold, and furnaces. The work also reveals an awareness of the serious cultural crisis brought on by a lack of interest in the study of things and a slow degeneration of language. "I have written only about things which I have seen, read, or carefully examined myself after learning about them from another." By contrast, Agricola was a harsh critic of the deliberate obscurity of language and the arbitrary character of the terminology used by alchemists, whose writings were "obscure" because their authors "use strange names [. . .] invented by themselves, though the thing itself changes not" (Agricola, 1563: 4–6 Preface).

In his 1556 commentary on Vitruvius, Daniele Barbaro pointedly asked "why have practical men not acquired credit? For the reason that architecture is born of discourse. Why have men of letters? For the reason that architecture is born of construction [. . .] To be an architect, which is an artificial generation, one must seek discourse and construction together" (Vitruvius, 1556: 9). The actual union between *discourse* and *practice*, *speculation* and *manufacture*, presented serious problems. For example, Bonaiuto Lorini, a military engineer for Cosimo de Medici in Venice, was perfectly aware of their importance. In a section from his 1597 treatise *Delle fortificationi*, he discussed the problem of the relationship between the work of the "purely speculative mathematician" and the "practical mechanic." The mathematician works with lines, surfaces and bodies that are "imaginary and separate from matter." His demonstrations "do not respond perfectly when applied to material things" because the type of matter that the mechanic deals with always has inherent "impediments." The mechanic's skill and judgment consists in being able to predict the problems deriving from the diversity of the materials he uses (Lorini, 1597: 72). This very problem of the relationship between "material imperfections" and "purely mathematical demonstrations" was addressed by Galileo in the introduction to *Discourses on Two New Sciences*.

A characteristic example of the "idealized" combined with the "physical," which makes repeated reference to Archimedes, is the work of Simon Stevin (Stevinus), born in 1548 in Bruges and died in 1620 in the Hague. He amazed contemporaries with a sailing chariot that he built for the amusement of the Prince of Orange and demonstrated on the beach at Scheveningen. Stevin wrote about arithmetic and geometry, worked on fortifications, designed and built machines and water mills, and created tables for calculating interest. *De Thiende* (The Tithe, 1585) was about decimal fractions and *De Havenvinding* (1599) about determining longitude. Stevin believed that Dutch was one of the oldest languages in the world and had the great quality of being more concise than any other. Because he addressed an audience of craftsman, he took pains to be very clear and so wrote in Dutch and not Latin. The title of the three-volume *Beghinselen der Weeghconst* (Elements of the art of weighing), published in 1586, refers to the medieval *scientia de ponderibus*. They were translated into Latin as the *Hypomnemata mathematica* (1605–8) and then translated into French in 1634.

Knowledge that can Grow

Two concepts emerge from the fifteenth-century work of artists and experimentalists, and the sixteenth-century treatises on mining, navigation, ballistics, and the art of fortification. As we have already discussed, one is a new view of manual labor and the cultural function of the mechanical arts. The other is the concept of knowledge as a progressive construction, or the accumulation of a series of increasingly more complex or more perfect results.

For this reason, technical knowledge was developing as a grand historical alternative to magic and alchemy and a typically Hermetic image of knowledge according to which scholars restated the same immutable truths over the course of thousands of years. In the Hermetic tradition, truth does not emerge over time, but is the perpetual revelation of an eternal *logos*. History only appears changeable and varied; in fact it contains a single immutable *sapientia*. The mechanics took a diametrically opposed view. In the introduction to *Diverse et artificiose macchine* (1588), Agostino Ramelli wrote that the mechanical arts developed out of the labor and need of the first men who sought to defend themselves in a hostile environment. Those arts developed not like the impetuous motion of the wind that could suddenly capsize a ship at sea and then die out, but instead like the course of a river that begins small and grows larger and mightier as it reaches the sea (Ramelli, 1588; Introduction). In his introduction to the *Treatise on the Proportions of the Human Body* (1528), Albrecht Dürer carefully explained why he, though not a scholar, dared address such an important subject. He decided to publish his book, despite the risk of criticism, for the public benefit of all artists and to encourage others to do the same "so that our successors may have something to perfect and advance" (Dürer, 1528: Dedication). Ambroise Paré, a self-taught surgeon who knew no Latin and was despised by his colleagues, declared that one should not rest on the laurels of the ancients because "there is more to discover than we know today and the arts are not so perfect that additions cannot be made to them" (Paré, 1840: I, 12–14).

The ideas that men such as Bacon, Descartes, and Boyle considered in a philosophical light were ideas that sprang from alien, non-philosophical milieus. In fact, their origins were considered with suspicion, if not outright contempt, by the academic culture.

Art and Nature

Though Bacon is really no longer called the "founder of modern science," it is certainly true that he brought to the fore ideas that had been marginalized by official science and that were drawn from the culture of builders and engineers such as Biringuccio and Agricola. Bacon respected the mechanical arts for three reasons: (1) they helped unveil the processes of nature, and are a form of knowledge; (2) they build on past experience and unlike other forms of traditional knowledge are cumulative, advancing so swiftly that "man's desires are satisfied before the arts can even be perfected"; (3) unlike other cultural forms, the mechanical arts permitted collaboration and are a form of collective knowledge in which "the wisdom of many meet, while in the liberal arts the wisdom of many is subjugated to that of a single person whose followers mainly corrupt it rather than further it."

The Baconian Robert Boyle (1627–91) felt that the book of nature, the artisan's workshop, and the anatomical room were strictly opposed to libraries, scholarly and humanistic studies, and purely theoretical research. His dispute in many cases verged on a sort of scientific primitivism. In *Considerations Touching the Usefulness of Experimental Natural Philosophy* (1671), Boyle clearly framed the interests and goals of Baconian groups. Experiments conducted by experts in laboratories were impressively accurate but artisans in workshops – while less accurate – compensated for this through their greater diligence. The title of the fourth essay in *Considerations* is very telling: *Touching the Usefulness of Experimental Natural Philosophy*.

The Baconian notion that the useful arts could shed light on theory was most clearly expressed by Gottfried Wilhelm Leibniz (1646–1716) with reference to the work of Galileo and Harvey. In the *Initia et specimina scientiae novae generalis pro instauratione et augmentis scientiarum ad publicam felicitatem*, Leibniz states that progress in the mechanical arts has largely been ignored by learned men. While technicians have no idea how their experiments can be utilized, scientists and theorists have no idea how many of their *desires* could be realized by the work of mechanics. The program of a history of the arts was energetically taken up in *Discours touchant la methode de la certitude et l'art d'inventer*: Knowledge that is not written down or codified and is scattered among men who do technical jobs of different sorts surpasses by far all the written knowledge housed in libraries. The greatest part of the treasure available to men has not yet been registered. Furthermore, there exists no mechanical art so small and "contemptible" that it cannot contribute observations and vital material for science. What is needed is a *theater of human life* based on human experience because the loss of even one of the arts could not be made up for by libraries. Leibniz believed that *writing down* the procedures followed by craftsmen and technicians was the most urgent task of the new culture.

In the preface to the *Encyclopédie ou dictionnaire raisonné des sciences, des arts et des mestiers* (1751), Jean D'Alembert (1717–82) reveals an awareness that the *Encyclopedie* was the fulfillment of a historical mission. In Chambers encyclopedia, wrote d'Alembert, one word was found for the liberal arts where pages were required, and in the mechanical arts everything needs to be started from scratch. Chambers had read books and never observed craftsmen at work, and there are some things that can only be learned in workshops.

In the *Prospectus* of 1750, Denis Diderot (1713–84) also expressed the importance of learning technical processes from direct observation: "We have turned to the most skilled workers in Paris and all of France, and have taken pains to visit their workshops, question them, write down what they dictated to us, flesh out their ideas, use the precise terms of their trade, compile them in tables, define them . . ." (Diderot, 1875–7: XIII, 140). In the article on *Art*, Diderot pointed to the harmful consequences that stemmed from the traditional distinction between the liberal and mechanical arts. A prejudice had developed against "turning to sensible and material objects" because it represented "a derogation of the dignity of the human spirit." This prejudice, he continued, "has filled the cities with proud reasoners and useless contemplatives and the countryside with petty, ignorant, lazy, and disdainful tyrants." The controversial defense of the mechanical arts was united to the larger theme of political equality.

Daedalus and the Labyrinth

Innumerable contemporary philosophers, commentators, and journalists have lumped all of modernity under the heading of a dangerous and unacceptable exaltation of technology and have identified Francis Bacon as the spiritual father of a "neutral technicalism" that lies at the heart of processes of alienation and commercialization so typical of modernity. The exact opposite is true. In all the many pages written about technology and its ambiguous nature, there are very few pages that resemble the Lord Chancellor's 1609 interpretation of the Daedalus myth. Daedalus was a clever but detestable man, best known for "illicit inventions" such as the machine that allowed Pasiphaë to mate with a bull and give birth to the Minotaur who devoured youths, and the Labyrinth he designed in which to hide the Minotaur and "protect evil with evil." One may conclude from the myth that the mechanical arts can produce useful objects and at the same time "instruments of vice and death." Technical knowledge, in Bacon's opinion, though capable of producing evil at the same time offered the possibility of diagnosing evil and curing it. Daedalus also created "solutions to his crimes." He supplied the thread that could penetrate the labyrinth: "He who devised the maze also demonstrated the need for the thread. The mechanical arts are indeed of ambiguous purpose and simultaneously produce evil and its remedy" (Bacon, 1975: 482–83).

The leaders of the Scientific Revolution believed that re-establishing human control over nature and the advancement of learning were of value only insofar as they were part of a larger context that involved religion, morality, and politics. Campanella's "universal theocracy," Bacon's "charity," Leibniz's "universal christianity," and Comenius's "universal peace" cannot be divorced from the interest and enthusiasm of these men for the new science, and formed arenas in which scientific and technical knowledge could function as tools of redemption and deliverance. For Bacon and Boyle, as for Galileo, Descartes, Kepler, Leibniz and Newton, human will and the desire to dominate were not the highest priorities. Nature was at the same time an object of domination and of reverence. It was to be "tortured" and bent to serve man, but it was also the "God's book" to be read with humility.

4

The Unseen World

Printing

Reading a book is an individual activity done quietly and on one's own. We take this act so for granted that it is hard to imagine that the familiar object in our hands was once such a shocking innovation that it not only made the communication of ideas and knowledge possible but also replaced the custom of reading unpunctuated texts aloud in a group (McLuhan, 1964). The printing press, gunpowder, and the compass are three mechanical inventions that are often grouped together. In *The City of the Sun* (1602), Campanella vividly describes them as a series of conquests that went hand-in-hand with the acceleration of history itself: "there has been more history in 100 years than in all the world in 4,000; and more books have been written in these 100 years than in 5,000; and the magnet, printing, and guns are great signs of world unity" (Campanella, 1941: 100). In 1620, Francis Bacon wrote that their effect was one of such radical change "that no empire, sect, or star seems to have exerted greater power in human affairs than these mechanical inventions" (Bacon, 1975: 635–6).

This was no overstatement. The fusion of different processes (the manufacture of paper and ink, the development of moveable and metal type, and printing itself) into a completely new technology introduced to Europe – 300 years early – the "theory of interchangeable parts" that lay at the heart of modern manufacturing technology (Steinberg, 1955). Johannes Gutenberg began printing books in Mainz (his Bible dates from 1456) using a technique that was perfected in the sixteenth century, remained unchanged until the nineteenth, and is even still used today. In 1480, there were active printing presses in more than 110 cities in Europe: 50 were in Italy, 30 in Germany, 8 each in Holland and Spain, 5 in

Belgium and Switzerland, 4 in England, 2 in Bohemia, and 1 in Poland. By 1500, only twenty years later, 286 cities boasted presses. L. Febvre and H. J. Martin have calculated that by 1500, 35,000 editions of 10,000–15,000 different works had been printed, and at least 20 million books were in circulation. Over the course of the seventeenth century there were 200 million copies in circulation (Febvre and Martin, 1958: 396–7).

The small-format books published by Aldo Manuzio have been compared to today's paperbacks. Along with Paris and Lyon, Venice became a major publishing center. At the end of the sixteenth century, the first international book fairs were held in Lyon, Medina del Campo, Leipzig, and Frankfurt. The number of copies per edition could vary from 300 to 3,000 but the average was 1,000.

The spread of ideas and the advancement of learning now required large capital investments and considerable risk-taking for the entrepreneur. These certainly had not been typical problems associated with scholarship for either the monk or the humanist.

Ancient Books

The great Italian humanists (Leonardo Bruni, Guarino Veronese, Giannozzo Manetti, and Lorenzo Valla, to name a few) believed that studying the great classics of the ancient world signified a return to a civilization superior to theirs, and the unattainable model for all forms of human society. Yet the humanists were not simply passive imitators, and their writings revealed a constant debate about the "barbarity" of medieval Scholasticism and the dangers of classicism itself. The need for *aemulatio* rather than *imitatio* became the battle cry of European intellectuals from Angelo Poliziano to Erasmus of Rotterdam. The lost texts that were rediscovered and commented on by humanists were anything but simple documents. The ancient texts upon which they exercised their refined philology contained – in their eyes – knowledge that was immediately useful to science and its practice. Access to books published in the original Greek, and to translations no longer based on medieval Arabic translations of Greek originals, had a significant effect on the development of scientific knowledge. Important editions included Euclid in the original (Basel 1533) and its Latin translation by Federico Commandino (Pesaro, 1572); Archimedes in Greek (Basel 1544) and Commandino's Latin translation of it (Venice 1558); as well as, once again, Commandino's translations of the *Conics* by Apollonius and the mathematical collections of Pappus (Bologna 1566; Pesaro 1588); Ptolemy's *Almagest* (Basel 1538) and translations of his *Geography* (virtually unknown until the Middle Ages). The first Latin translation of Hippocrates (Rome 1525) was quickly followed by its publication in the original Greek in 1526 (Venice) and 1538 (Basel). Galen's vast opus (much of which had been translated in the Middle Ages from Arabic and contained many spurious inscriptions) was carefully organized and integrated after the discovery of treatises previously unknown in the Western world. Galen's work appeared first in Latin in 1490 (Venice) and in Greek in 1525 (Venice), and was followed by the edition edited by Joachim Camerarius and Leonhart Fuchs (Basel 1538).

The Old and the New

The relationship between two characteristics of the Renaissance – the *rediscovery of the ancients* and *the sense of the new* – were related in a complex way because the leaders of the Scientific Revolution regarded antiquity with a completely different attitude than had the humanists. Bacon and Descartes used classical texts while at the same time rejecting the exemplary nature of the classical world. They rejected not only pedantic imitation and passive repetition but also *aemulatio*, still considered essential by humanists but not so by these new thinkers. Indeed even the humanist contextual basis for comparison with the ancients was rejected: Descartes pointed out that when one spends too much time traveling one becomes a foreigner at home, much in the way that one who is too curious about the past becomes ignorant of the present. Bacon felt that the spirit of the ancient Greeks was narrow and limited, and if their path were followed it would be impossible to imitate them. It was necessary to change course and take on the role of "not of judges but of guides" (Bacon, 1887–92: III, 572).

In 1647, Blaise Pascal still believed that it was not possible to propose new ideas with impunity because respect for antiquity "had reached such a level that all its thoughts are taken as oracles and all that is unintelligible is considered a mystery" (Pascal, 1959: 3). Yet even *aemulatio* no longer made sense. The ancients explained the Milky Way in the only way they could given their limited resource: the naked eye. The fact that we know more about nature today than they did allows us to hold new opinions without offending them or seeming ungrateful. For this reason, and without contradicting them, we may say the opposite of what they said (ibid.: 7–8, 9–11).

Because the new astronomy not only pushed the outer reaches of the universe immeasurably but in some cases even proclaimed it was infinite, it gave many the distinct impression that traditional knowledge as they knew it had come to an end. In 1657 Pierre Borel wrote that "we know nothing that is not or cannot be debated." Astronomy, physics, and medicine "vacillate every day and see their very foundations crumble." Petrus Ramus had destroyed Aristotle's philosophy, Copernicus Ptolemy's astronomy, Paracelsus Galenic medicine: "we are compelled to admit that what we know is much less than what we do not know" (Borel, 1657: 3–4).

The awareness that knowledge was on the brink of great change was a notion that could excite and inspire, but even more frequently amaze, bewilder or frighten, and one finds confirmation of this in many documents of the time. Is it not clear, wrote John Dryden, that in this last century a *new* nature has been revealed to us? The theme of novelty was pervasive in all of European culture and the word *new* appeared almost obsessively in the titles of hundreds of scientific publications in the seventeenth century: *Novum Organum* by Bacon, *Nova de universis philosphia* by Francesco Patrizi (1591), *De mundo nostro sublunari philosophia nova* by Gilbert (1651), *New Astronomy* by Kepler (1609), *Discourses on Two New Sciences* by Galileo (1638), *Novo teatro di machine* by Vittorio Zonca (1607), to name just a few (Thorndike, 1957).

Illustration

According to Panofsky (who published a major monograph on Dürer in 1943), the rigorous depiction of the natural world in the great paintings and engravings of the fifteenth to seventeenth centuries had the same importance for the descriptive sciences as the invention of the telescope and microscope had for astronomy and the life sciences. The illustrations in botanical, anatomical, and zoological texts were not simply additions to text but valuable given the absence of an adequate technical vocabulary (only achieved in botany, for example, over the course of the nineteenth century). Artistic collaboration in the descriptive sciences had a revolutionary effect.

For this reason let us return to Leonardo and his painterly interest in rendering all things *visible*. Many of his drawings of rocks, plants, animals, clouds, and the movement of air and water reveal scientific knowledge of living nature. Leonardo's anatomical drawings changed significantly after 1506, which is probably when he read Galen's *On the Use of Parts* and when the practice of dissection was becoming more common. For many years Leonardo studied in great depth the comparative anatomy of vertebrates, the flight of birds, and physiological optics, and made endless drawings of these subjects. He studied horse anatomy and made hundreds of drawings for the projected equestrian statue of Ludovico Sforza, the Duke of Milan (on which he began working in 1483) and the grand painting of the *Battle of Anghiari* (begun in 1503).

But Leonardo's curiosity was far greater than that of other sculptors and painters interested in artistic or superficial anatomy. He was a methodical and systematic observer and this approach was tied to his belief in the superiority of the *eye* to the *mind*, of careful observation of the real world to books and treatises. This was the source of both his greatest limitation (often cited by those who object to the mythic image of Leonardo the "modern scientist") and his one-of-a-kind greatness.

Leonardo's drawings remained unknown. The first woodcuts used to illustrate books printed with mobile characters date to 1461. The transition to engraving (Dürer's work among the most celebrated) and etching (a technique used by Rembrandt) led to a progressive refinement of the art of illustration. The first illustrated text on anatomy was Berengario de Carpi's 1521 commentary on Mondino de' Luzzi's *Anatomia* (de Luzzi was a professor in Bologna between 1315 and 1318). This was followed by *Isagoges breves in anatomiam* in 1523, and one of the greatest examples of the genre, *De dissectione partium corporis humani* (1545) by Charles Estienne (Stephanus Riverius). But the large and exquisite anatomical engravings by Andreas Vesalius in *De humani corporis fabrica* surpass all others for their precision and accuracy in depicting human anatomy and justifiably became the symbol of a radical change in the observation of the real world. Vasari attributed them to Jan Stephan van Calcar who was a student of Titian. One need only compare Vesalius's illustrations to medieval anatomical drawings to appreciate the quantitative leap they represented in the observation and representation of the human body. It is often observed that in 1543 Copernicus gave men a new image of the universe and Vesalius a new view of their bodies. Vesalius, who was born in Brussels to a family of doctors, studied in Louvain and Paris, and spent a great deal of time in Italy; he was called to Padua in 1537 to teach anatomy, and subsequently also taught in Bologna. In 1538 he published six anatomical sheets known as *Tabulae sex*. In 1543 he went to Basel to personally supervise the printing of the *De Fabrica* and the *Epitome* (also published in that year). He was only 28 years old when his masterpieces were published and he wrote in the Preface, "I am aware that because of my youth my work will carry little authority and be criticized because its Galenic principles do not correspond to reality [. . .] unless it finds an illustrious protector." That illustrious protector was none other than Emperor Charles V, who appointed Vesalius imperial physician and to whom the book is dedicated.

Vesalius patterned the content of the sections after Galen, and reaffirmed Galen's belief that the venous system was more important than the arterial to circulation. Like Galen, he too believed the veins originated in the liver. However, he energetically stated in the preface that he was distancing himself from the Galenic tradition because the master "was unaware of the numerous and substantial differences between the bodies of monkeys and men, with the exception of how differently the toes and ankles bend"; and that Galen, in the course of a single anatomical demonstration, "erred more than 200 times in correctly describing the parts, the harmony, the use and the function of the human body."

Contemporary scholars who have emphasized Vesalius's "Galenism" have not only tended to overlook these statements but also have not taken into account how fiercely the *De Fabrica* was criticized by traditional Galenists. Jacques Dubois (Jacobus Sylvius), Vesalius's teacher in Paris, became his greatest adversary and enemy, constantly referred to him as *Vesanus*, a play on words that meant "crazy or delirious," and accused him of poisoning the world of medicine with his work. Vesalius enthusiastically endorsed the need to fully merge clinical medicine with dissection (and surgery); he criticized the bookishness of medicine and fought for the union of direct observation and theory in the field. He proposed a new ideal of the physician, the professor of medicine, and the relationship in the "experimental" sciences between manual and intellectual labor. He felt that "disdain for manual labor" was one of the reasons for the degeneration of medical science. Doctors limited their services to prescribing medicinals and diets and had thrown aside the rest of the discipline to those "whom they call surgeons and consider nothing more than slaves." Once all manual procedures had been turned over to barbers, "not only did doctors lose their knowledge of internal organs but their cutting skills were also diminished." Doctors would not personally risk performing operations, yet those who did perform them were too ignorant to understand the written work of doctors. And so a reprehensible custom was established, with one person performing the operation and another describing the parts. The latter "presumptuously jabber from on high" and monotonously repeat things never directly observed but only memorized from books. Instruction was so poor that students learned less in that chaotic atmosphere than a doctor could have learned from a butcher in his shop" (Vesalius, 1964: 19, 25, 27). In 1555 the second edition of *De Fabrica* was published with some minor corrections. Vesalius was appointed physician to Philip II of Spain. He gave up the post in 1562 and died two years later in a shipwreck while returning from a pilgrimage to the Holy Land. He was returning to Padua to teach upon appointment by the Venetian Senate.

Vesalius's great book was also visible proof of an ever-growing collaboration between natural scientists and artists and engravers. Studies of illustration techniques and forms of this problematic collaboration in the field of anatomy as opposed to engineering, zoology, anatomy and botany, have emphasized the rapid and extraordinary transition that occurred in the sixteenth century from text-based to nature-based illustration. Two important books signaled the beginning of the tradition of modern herbals: *Herbarum vivae icones* (1530–6) by Otto Brunfels and illustrated by Hans Weiditz, and *De historia stirpium* (1542) by Leonhart Fuchs. In each case their novelty stems more from the illustrations than the text. In the preface to *Historia,* Fuchs wrote that the greatest care was taken "to represent every plant with its roots, stele, leaves, flowers, seeds, and fruit; moreover the omission of any modification of the natural form of a plant using shadows or other artistic techniques has been deliberate." At least in this case, there was some supervision; "we have not allowed artists to indulge their whims in order to avoid that the representations not correspond precisely to reality" (Fuchs, 1542: Preface). The first university botanical gardens were established in Padua and Pisa in 1544. In the early seventeenth century, "gardens" were second only to anatomical theaters as indispensable elements to insure the respectability of a university.

Far fewer encyclopedic works in zoology were undertaken. Some important "specialized" histories of animals (with important illustrations as well) include: *La nature et diversité des poissons* (1555) and *L'histoire de la nature des oyseaux* (1555) by Pierre Belon; *De piscibus marinis* (1554) by Guillaume Rondelet and *Dell'anatomia et delle infermitadi del cavallo*, a splendid treatise by the Bolognese senator Carlo Ruini. In addition to the work of Ulysses Aldrovandi, a major Renaissance contribution to the field was Konrad Gesner's *History of animals*. Though he died young, Gesner – a doctor and humanist – published books on botany, linguistics, the Alps, and alpinism. In 1545, at the age of 29, he published *Bibliotheca universalis*, a compendium of titles in Latin, Greek, and Hebrew. The *History* was published in five folio volumes and three volumes of *Icones* between 1551 and 1558 (the fifth volume was published posthumously in 1587) and numbered roughly 4,500 pages with over 1,000 woodcut illustrations made by artists in Gesner's native Zurich. The famous picture of a rhinoceros – to remain the model for all illustrations of rhinoceroses until the end of the eighteenth century – was based on Dürer, and reveals that artist's idea of the most famous of "exotic" animals as a scaly dragon (Gombrich, 1960). An extra small spiral horn behind the ears near the cervical vertebrae vanished from illustrations only in 1698.

Gesner ignored comparative anatomy and classified animals alphabetically (hence *Hippopotamus* comes between *Hippocampus* and *Hirudo* or leech). Each animal was described in often lengthy chapters (176 pages in folio for the horse and 33 for the elephant) that were divided in sections designated by letter. The sections included the animal's name in different ancient and modern languages, its habitat and morphology, diseases, behavior, function and breeding, edibility (when appropriate), usefulness to medicine, etymology, and proverbs.

On the subject of illustration and the "limitations of realistic portrayal," Gombrich certainly is correct in sustaining that a representation that already exists "will always exert an influence on the artist even when his goal is to accurately depict reality" and "a visual image cannot be created from nothing." Moreover, the comparison of a drawing of a lion and a porcupine by the Gothic architect Villard de Honnecourt and a Dürer watercolor of a rabbit reveals something that Gombrich too has pointed out: between the fourteenth and the sixteenth centuries a significant stylistic change occurred. Artistic "style" lost its formality and "learned to adapt with sufficient ease" to the subjects it rendered (Gombrich, 1950). This change had a significant impact on the development of scientific knowledge.

New Stars

In 1609 Galileo Galilei began observing the night sky with a telescope and then published his observations in Venice, on March 12, 1610, in a short book entitled *Sidereus nuncius* (*Starry Messenger*). Galileo described the surface of the moon as not "uniformly smooth and perfectly spherical as countless philosophers have claimed about it and other celestial bodies, but rather, uneven, rough, and full of sunken and raised areas like the valleys and mountains that cover the Earth." Looking more closely, he observed that the boundaries between dark and light regions were uneven and sinuous, and points of light that appeared in the shadowy region of the moon merged with the light regions. Didn't the same thing happen on Earth? The highest mountain peaks are illuminated by the dawn's light while the plains below remain in the shadow. And once the sun has risen, the sunny mountains ultimately join together with the plains. The landscape of the moon is therefore *like* the landscape of the earth. The Earth has characteristics which are *not* unique in the universe. The celestial bodies, at least in the case of the moon, are not by their nature different and do not possess the characteristics of absolute perfection that an age-old tradition attributed to them. And there are many more stars than can be seen with the "naked eye." The telescope revealed a sky populated by countless heavenly bodies. It showed the complex structure of known constellations, and uncovered the nature of the Milky Way: "what we observed in the third place is the essence, or rather the matter of the Milky Way that, thanks to the telescope, has been carefully examined, and the mysteries that have plagued philosophers for centuries can now be solved by virtue of what we are able to see with our eyes and so are freed from verbose discussion." The ability to see the dark sections of the moon's surface led Galileo to conclude that sunlight was reflected from the Earth to the moon and back. As for the stars, he finally concluded that there was a great difference between fixed stars and planets: stars still looked like bright points surrounded by "sparkling rays" but did not appear larger through the lens of the telescope, whereas planets looked like perfectly round and definitely bounded globes, like little moons and were much magnified by the telescope. So the distance between the fixed stars and Earth was incomparably greater than that between the planets and Earth.

Galileo also presented another fundamental discovery in the *Sidereus nuncius*, the pages of which still today resonate with the sense of trepidation that accompanies any vision of a new truth. On the night of 7 January he observed three small bright stars next to Jupiter, two to the east and one to the west of it. The following night their position had changed and they were all west of Jupiter, and on the 10th two of the stars were to the east and the third seemed hidden behind the planet. On the 12th, after two hours of observation, Galileo watched the third star reappear, and on the next night there were four stars: these were Jupiter's moons or satellites (now known as Io, Europe, Ganymede, and Callisto) and called the "Medicean stars" by Galileo in honor of Cosimo II.

The revolutionary nature of Galileo's discoveries was recognized by his contemporaries. In a poem dedicated to "our century's prince of mathematicians," Johannes Faber declared that Vespucci and Columbus who had sailed uncharted seas should make way for Galileo who had given all of mankind new constellations. The comparison to great geographical discoveries and New World voyages was repeatedly made. In England, William Lower wrote to his friend, Thomas Hariot, that Galileo had accomplished something even greater than Magellan who had also opened up unexplored territory. In a 1612 work describing the intellectual world in which he lived, Francis Bacon praised "the industry of mechanics, the zeal and energy of certain learned men who so recently, by using new optical instruments like sloops and small vessels, had begun new enterprises with the wonders of the heavens." Their undertaking, he continued, should be regarded as "something noble and worthy of mankind and those brave individuals should also be praised for their honesty because with candor and clarity, they methodically accounted for each step in their research." Though the Lord Chancellor did not accept Copernican cosmology, he was a great philosopher. The English ambassador to Venice, Sir Henry Wotton, though a man of vast and refined culture, was not. The day *Sidereus nuncius* was published he sent a copy to the king with a promise to soon send him a telescope and his words clearly conveyed the challenge that Galileo's observations posed to traditional cosmology: "I send His Majesty, together with this letter, the strangest news

ever before heard on earth. By this I mean the enclosed book by that
mathematics professor at Padua [. . .]. He hath first overthrown all former
astronomy [. . .] and next all astrology [. . .]. The author runneth a for-
tune to be either exceedingly famous or exceedingly ridiculous."

There was no shortage of bitter controversy, tenacious rejections, ob-
stinate shows of disbelief from pro-Aristotelian academic quarters. The
celebrated Cremonini, a colleague and friend of Galileo at Padua, did not
believe Galileo had seen anything, objected to those "spectacles" that
"muddle one's head," and berated Galileo for having gotten involved
"in all these intrigues." Giovanni Antonio Magini, an astronomer at
Bologna, took a hostile and malicious stance. And when Galileo was in
Bologna in April 1610 to try to prove his discoveries to the faculty, Martin Horky, who later became a vehement enemy, wrote to Kepler saying
that he "had tested Galileo's instrument in many ways, with both inferior and superior things; while remarkable with the former it fails with
regard to the heavens because the fixed stars appear exactly the same."

Eventually Kepler acknowledged Galileo's work, and after initial
skepticism so did the Jesuits in Rome. Galileo had won, because in order
to win over the final, unyielding objectors and silence the professors who
rejected the moon's mountains or Jupiter's satellites on the grounds of
logic and metaphysics, not even "the sworn testimony of the stars themselves" would have sufficed, as he later wrote. The reality of the universe
had been expanded through the use of a mechanical instrument that was
capable of assisting, perfecting, and refining the human senses. For the
men of his day, Galileo's astronomical observations not only signaled
the end of one world view but represented the birth of a new concept of
experience and truth. "Seeing is believing" and the evidence had at last
broken an endless circle of dispute.

The Discovery of an Unknown Visual World

In the seventeenth and eighteenth centuries there was as great a fascination with the *small* and infinitely small as with the *large*, such as endless
distances and an apparently infinite universe. The concept of nature as a
sort of *plenum formarum*, an infinite hierarchy of forms, a complete and
endlessly graduated scale of Being (one of the great idea-forces in philosophy for the last two centuries) seemed in itself to imply the existence
of a minute and invisible reality which necessarily escaped the limited
capacity of the human eye. In 1664 Henry Power published *Experimental Philosophy, Containing, New Experiments Microscopical, Mercurical,
Magnetical*. He believed that "new dioptrick discoveries" were proof
that the smallest bodies man could see with the naked eye were but "a

proportional average" of two undetectable extremes. Even the idea that nature could be explained by studying its corpuscular or cellular structure implied an interest in instruments that could magnify the natural range of man's senses. The inhabitants of Bacon's *New Atlantis* (1627) had visual aids superior to the spectacles and lenses in use at the time and that allowed them to "see small and minute bodies perfectly and distinctly, such as the shapes and colors of small flies and worms, grains and flaws in gems, and observations of urine and blood, that cannot otherwise be seen" (Bacon, 1975: 861).

Yet in the history of the microscope and its contribution to science there is no single dramatic date that compares to 1609 in the history of the telescope. It has often been noted that the telescope exerted its influence within the framework of an established science deeply rooted in antiquity. The microscope, on the other hand, was introduced at the beginning of a long process that culminated in new sciences. Histology and microbiology were only established in the eighteenth century. The word *microscopium* was first used in a letter from Johannes Faber to Prince Federico Cesi (April 13, 1625). Cesi and three friends had begun to study science in 1603 as young men, and were the nucleus of the *Accademia dei Lincei*. The first "separate" book on microscopy was the *Centuria observationum microscopicarum* (1655) by Pierre Borel.

The early seventeenth-century microscope was a tubular instrument with a lens at one end and a place to put an object on a glass slide at the other. Objects were magnified about ten diameters. The first members of the *Lincei* used instruments of this kind, and indeed the name of their academy refers to the acute eyesight of the lynx. In 1625 Cesi added plates of bees to the *Apiarum*, and these were published in Stelluti's *Persio tradotto* (Rome, 1630). This was most likely the very first printed illustration of objects seen with the use of a microscope. Stelluti insisted that what could be seen in the plates "was unknown to Aristotle and every other naturalist." In addition to an image of a "walking" bee, there were illustrations of its wings, tongue, "hairy eye," and legs. In 1644 in Palermo, Odierna studied the compound eye of different insect species. Two years later in Naples, Fontana completed a study of vinegar eels.

The so-called "classical microscopists" – Robert Hooke, Antony van Leeuwenhoeck, Jan Swammerdam, Marcello Malpighi, and Nehemia Grew – belonged to the next generation. The instruments they used could enlarge objects up to 100 times, though resolution was mediocre. The compound microscope (not used by Leeuwenhoeck) consisted of lenses fixed at the ends of cardboard tubes. The tube of the eyepiece was fitted inside that of the objective and to focus one slid the tubes up and down. Microscopes of this type (made in Italy by Campani) were widely known. The type described by Hooke had a screw-mechanism for focusing and was made up of a large cylindrical body: the objective had a biconvex lens regulated by a diaphragm and the eyepiece was made up of a flat-convex lens and a small biconvex lens (the mirror that cast light upwards was not invented until about 1720). These microscopes (as well as Leeuwenhoeck's incredibly minute lenses) were not limited to magnifying the known world (like the bees enlarged by Cesi) but opened up a new and unexpected world of minerals and organic tissue that had structures; a world inhabited by living organisms invisible to the naked eye.

Let us briefly return to the subject of illustration and its importance to science. Like Vesalius's work one century earlier, it was the fine engravings made by the celebrated architect Christopher Wren for Hooke's *Microscopia* (1665) that set the work apart from others of the same period. And Hooke's contemporaries included Marcello Malpighi, the author of *De pulmonibus* (1661), who was certainly a better *biologist* than Hooke. For a century and a half it had been acknowledged that the art of illustration had much to offer the sciences, yet the first generation of microscopists seemed hardly to notice. The 32 fine plates in the *Microscopia* (still used in nineteenth-century manuals) illustrated the potential of the medium (Hall, 1954).

The point of a sharp needle, fleas, flies, ants, lice: Hooke not only observed objects never before observed by others, but he described what he saw with unusual precision and passion for detail. He described the outer covering of a fly's eye as supple and transparent, and like the human cornea. When the eyeball and dark, mucousy substance was removed Hooke noted "how transparent the covering is, like a thin layer of skin with many small holes in the same arrangement as the external prominences." It was clear that this curious apparatus was the organ of sight for flies and crustaceans (Hooke, 1665: Preface). The term *cell* was first used in the title of his eighteenth observation, "Of the Schematisme or texture of cork, and of the cells and pores of some other such frothy bodies." He used the term because of the resemblance to the cells of the honey-comb, and so it is a mistake to consider him the discoverer of the cell.

Hooke, a scientist in the Baconian tradition, often emphasized the notion of broadening the territory of the senses. The telescope had opened up the skies to observation and revealed "a great number of new stars and new movements unknown to ancient astronomers." At the same time the once familiar Earth now seemed new, and every observation of its matter revealed "as great a variety of organisms as once we counted in all of the universe." New instruments permitted both the examination of the visible world and the discovery of unknown worlds: every notable improvement of the telescope and microscope "produces new worlds and lands unknown to our sight" (ibid.: 177–8).

Over the course of several meetings of the Royal Society in 1677, Hooke read from a 17–page letter by Antony van Leeuwenhoeck to that illustrious academy. The author was neither a natural philosopher nor a scholar. Leeuwenhoeck, chamberlain to the aldermen of Delft in southern Holland, had single-handedly manufactured hundreds of tiny biconvex lenses (smaller than 2.5 mm. in diameter) which, placed in a metal mounting, worked as simple microscopes. His remarkable skill as an optician (one of the lenses he developed proved to be superior to any other known simple lens) together with unsatiable curiosity led him to examine spermatozoa and red blood cells and identify protozoa and bacteria. In September 1674 he described the movement of the organisms in a drop of water [. . .] swift and marvelous to see and I judge that some of these little creatures were above a thousand times smaller than the smallest ones I have hitherto seen in cheese or mould." Small animals even live inside the human body. In October 1676 he described protozoa: "it is like seeing, with the naked eye, little eels that twist and turn against each other and the water seems to be alive with these different little animals; this for me, of all the marvels I have observed in nature, is the grandest of all."

The New World

"In the Indies everything is marvelous, surprising, different, and on a larger scale than anything in the Old World," wrote José Acosta. Columbus, Magellan, and many other early modern explorers and navigators, had all seen with their own eyes – like Galileo, Hooke, and Leeuwenhoeck did later – things never before seen. The discovery of new lands also challenged the superiority of antiquity. Simple mariners – declared many – were able to *see* the opposite of what Greek philosophers and Church Fathers had declared with regard to the inhabitability of tropical areas, the existence of the Antipodes, the navigation of the Oceans, and the impassability of the Pillars of Hercules.

The New World had unknown plants (corn, manioc, potatoes, beans, tomatoes, peppers, squash, avocado, pineapple, cocoa, tobacco, rubber trees) and unknown animals (turkey, llama, lynx, puma, condor, jaguar, tapir, vicuña, cayman). There were descriptions of new animals and plants in *Historia general y natural de las Indias* by Gonzalo Fernández de Oviedo y Valdés, who for over forty years was the gold inspector in Santo Domingo. In early sixteenth-century charts and maps, the new continent was shown to be inhabited by unicorns, dog-headed beings, and men with eyes, noses, and mouths on their chests: Oviedo refrained from such descriptions of monsters and imaginary creatures. He believed in one nature that assumed different forms in different parts of the world; plants that are dangerous in one location are not in another, man can have white or black skin, tigers that are agile and swift in Europe are "heavy and slow in His Majesty's India." Even the Jesuit Acosta described the characteristics of soil, minerals, volcanoes, metals, plants, animals, fish, and birds in the *Historia natural y moral de las Indias* (1590). The New World is inhabited by "many animals never before seen by the Greeks, the Latins, or any other people of *this world*." The great mathematician Thomas Hariot, who admired Galileo and corresponded with Kepler, touched on a similar theme in a 1588 pamphlet entitled *A Briefe and Troue Report of the New Found Land*. In Italy, Cesi acquired the manuscript of *Tesoro messicano* or *Rerum medicarum Novae Hispaniae thesaurus*, a collection of exotic plants and animals based on the report of King Philip II's physician, Francisco Hernández. After a series of editorial mishaps, Stelluti published the work in 1651.

Acosta also discussed at great length the men and women of the New World and their customs. The book, translated into English in 1604, Italian in 1606, and Dutch in 1624, was at the center of a great cultural debate in Europe from the middle of the sixteenth century to Vico. Many difficult questions were posed, such as how to reconcile Bibilical narrative with the fact that men lived in places remote from the center of the Judeo-Christian world; whether the indigenous people of the Americas descended from once-civilized races that had fallen into barbarism; did the different races and peoples in different parts of the Earth have different origins; how can one explain that all men are directly descended from Adam; did the Great Flood affect all the lands on earth, or was it a local disaster, and if local then is the Biblical account confined to one population in one location only; how can one explain the existence of plants and animals so different from those known; how did the New World animals get on and off Noah's Ark; why are there no survivors of these species in the Old World; is it possible that after the sixth day of creation God went on to create the New World; and most importantly, how did Old World peoples reach the New World?

Freethinkers of every sort used the discovery of the New World to express their doubts about the validity of the Biblical account and advance the type of sacrilegious questions that were popular in the late seventeenth and eighteenth centuries, calling them Lucretian, Spinozoan, and materialist. Cardano implicitly asserted the spontaneous generation of humans from matter. The Aristotelian Andrea Cesalpino explicitly stated that "all animals, including man, may have originated from decaying matter," and in his opinion, this was more easily seen in torrid and lush zones like the New World. Giordano Bruno believed that the presence of animals and men in the New World did not constitute a problem and was proof that each land produced its own kind of animals. It was absurd to assume that the peoples of the Americas were descended from Adam, and "indeed, there was not just one first wolf or lion or ox that begot all the wolves, lions, and oxen that live all over the world, but different lands produced their own kind from the origin." The debate between polygenists and monogenists (such as Acosta) was destined to be sensational.

Paracelsus would not attribute human characteristics to New World populations. Like giants, gnomes, and nymphs, "they resemble man in every way except for their souls." Like "bees with their king and wild ducks with their leader, they live not according to the order of human laws but follow their innate nature." Even the humanist Juan Ginés de Sépulveda, among other writers, philosophers and travelers, portrayed indigenous Americans as a subspecies capable of every sort of "abominable infamy." Montaigne presented a radically different view of Brazilian tribesmen in the *Essays* (1580): "it is neither possible nor just to take the point-of-view of Europeans or Christians when judging non-European peoples. Humanity expresses itself in an infinite variety of forms and "each claims the other uncivilized" (Montaigne, 1970: 272).

Discussion of "good and wicked savages" was interwoven with events in biology and political philosophy. Up until Buffon, the abbot Corneille de Pauw, and the Romantics, there was no doubt that all *nature* in the New World was "degenerate", "decadent", or even "inferior." Hegel described New World fauna in *Philosophie der Geschichte* as smaller, weaker, and less virile.

5
A New Universe

Copernicus

A Polish astronomer named Niklas Koppernigk (1473–1543) chose to Latinize his name to *Copernicus*. In the modern world, this name came to stand for a great development in thinking, the birth of a new age, and an intellectual revolution. Yet scholars have consistently pointed out that Nicholas Copernicus never took a revolutionary stance either personally or professionally. Like all good humanists, he believed that a new method of calculating the motions of the heavenly bodies (which would resolve unanswered astronomical questions) could be found in the work of the ancient philosophers. He presented his work as an attempt to revive the ancient doctrines of Pythagoras and Philolaus. Cautious and hesitant, he worried that his strange and unusual idea that the Earth moves would be met with "scorn" from clergy and scholars. His greatest work, *De revolutionibus orbium coelestium* (1543), was written as a parallel of Ptolemy's *Almagest* and follows its structure book by book, section by section. Kepler later remarked that Copernicus had interpreted Ptolemy rather than nature.

Copernicus was born in Toruń (Thorn in German), in a town on the river Vistula that had come under the rule of King of Poland in 1466. After his merchant father died, he was adopted by his maternal uncle (who later became the bishop of Ermland). He studied first at the University of Cracow, and then went to Italy at the urging of his uncle. According to the records of the *Natio Germanorum*, he was enrolled at the University of Bologna in 1496, where he was both the friend and pupil of the astronomer Domenico Maria Novara (1454–1504). In 1500 he went to Rome and the next year returned to Poland to be installed as canon of the cathedral of Frauenburg. He returned to Italy, however, that same year, first studying medicine and law at Padua for four years and then earning a doctorate in canon law from the University of Ferrara. In 1506, after nine years spent almost entirely in Italy, he returned to Poland to work as his uncle's secretary and physician. When his uncle died in 1512, Copernicus returned to Frauenburg where he lived and worked on his masterpiece for the rest of his life.

Between 1507 and 1512 (though the experts disagree on these dates) Copernicus drafted *De hypothesibus motuum coelestium commentariolus*, which circulated widely in manuscript form. The text presented the seven *petitiones* that would establish a new astronomy.

1   There is no one center of motion for all the heavenly bodies. The implication was that there were two centers of rotation – the planets around the sun and the moon around the Earth – instead of the one described by Ptolemy.
2   The center of the Earth is not the center of the universe, but only the center of gravity and the lunar sphere (this claim led to the need for explaining gravity).
3   All the planets revolve around the sun (which is therefore eccentric with respect to the center of the universe).
4   The distance from the Earth to the sun must be negligibly small compared to the distance of the sun from the sphere of fixed stars. This implies that the universe is so large that the movement of the Earth will not result in any apparent relative movement of the fixed stars.
5   All visible movement in the heavens results not from movements occurring in the heavens but from the movement of the Earth. The vault of the heavens is immobile, while the Earth and its closest elements (its waters and atmosphere) makes a complete diurnal rotation about its fixed poles.
6   The apparent movement of the sun results not from its own motion but from the Earth's motion around the sun. So, the Earth has more than one motion.
7   The apparent retrograde and direct motion of the planets results not from their motion but from the movement of the Earth. The Earth's movements are sufficient explanation for all apparent discrepancies in the sky (the so-called planetary "retrogradations" were actually *apparent movements* because they were dependent on the motion of the Earth).

At about this time, Copernicus handed over the lengthy manuscript of *De revolutionibus* to the young Georg Joachim Rheticus (1514–76), an early disciple and admirer. Rheticus too had changed his name, from

Lauschen in acknowledgment of his origins in the Roman province of Rhaetia. He published the *Narratio Prima* in 1540, a text that combined astrological reflections on the fall of the Roman Empire, the birth of the Moslem Empire, and Christ's second coming with a clear explanation of the Copernican theory of astronomy. With the book's second printing in Basel the following year, knowledge of Copernicus' work and ideas spread to a wider learned audience.

Rheticus enthusiastically claimed that the Copernican system was simpler and more harmonious than the Ptolemaic one. All planetary motion could be explained by the uniform motion of the Earth. By fixing the sun at the center of the universe and allowing the Earth to rotate around it in an eccentric orbit, the true intelligence of the cosmos relied exclusively upon the regular and uniform motion of the *planet Earth*. Why shouldn't Copernicus adopt the "convenient theory" of terrestrial motion? By doing so, in order to construct an exact science of heavenly phenomena one simply needed "the eighth fixed sphere, the sun remaining immobile at the center of the universe, and to explain the motion of the other planets one needed only combinations of eccentrics and epicycles, eccentrics and eccentrics, and epicycles and epicycles" (Rheticus, 1541: 460–1). Attributing motion to the planet Earth allowed for the reconfirmation that heavenly motion was circular in nature. Traditional cosmology accounted for backward motion by placing the planet in an epicycle the center of which rotated in turn around the Earth on the deferent of the planet. In the new system the planets were in constant motion, all in the same direction. Irregular movements were due to the ever-changing perspective of the observer standing on a moving Earth.

Tradition has it that the text of *De revolutionibus* (published in May 1543) was brought to Copernicus on his death bed. In the dedication, Copernicus – like Rheticus before him – insisted on the greater simplicity and harmony of his system. He compared new to old and pointed out the disagreements, doubts, and contradictions in the beliefs of traditional astronomers.

The Copernican revolution did not consist of an improvement of astronomical method nor in the discovery of new data but in the construction of a cosmology based on the *very same facts* provided by Ptolemaic astronomy. This view of the universe was moreover deeply tied to several basic Aristotelian themes: Copernicus' universe was perfectly spherical and finite; the roundness that all bodies tend to is a perfect form, and is a self-contained whole justly attributed to divine bodies; the circular motion of the crystalline spheres derives from the fact that their very mobility is circular (*mobilitas sphaerae est in circulum volvi*), the immobile condition of the sun (like the immobility of the fixed stars) springs from its divine nature and its centrality from the fact that this "lantern of the world" is located in the best possible place from which to "simultane-ously illuminate all things" (Copernicus, 1979: 212–13).

The new system seemed simpler than it actually was: in order to justify the observational data, Copernicus first had to designate the center of the universe as the center of the Earth's orbit, and not the sun (his system was described as *heliostatic* rather than *heliocentric*); then, like Ptolemy, he had to introduce a series of rings rotating around other rings; and finally he had to give the Earth (in addition to rotating on its own axis and around the sun) a third movement of declination (*declinationis motus*) in order to justify the invariability of the earth's axis with respect to the sphere of the fixed stars.

The Copernican revolution did not limit itself to challenging old ideas with new ones but succeeded in actually replacing Ptolemy and improv-ing upon the *Almagest* in terms of calculations and planetary tables. The new tables – drawn up by Erasmus Rheinhold (1511–53) and called the *Prutenic Tables* – were based on Copernican ideas and accepted by even the most vehement opponents of the new world system; even Rheinhold himself was never a Copernican. The system that *De Revolutionibus* de-scribed was based on sophisticated Pythagorean mathematics and im-pressed professional astronomers. Some of them found the new system not only simpler and more harmonious than the old one, but also more consistent with the metaphysical assumption (upheld by Copernicus) that celestial movement was perfectly circular.

Copernicus's masterpiece did not even include many of the basic ele-ments considered to make up the great phenomenon we call the "astro-nomical revolution": the elimination of eccentrics, epicycles, and the reality of solid spheres; or the suggestion of an infinite universe. Yet texts do exist which, though not revolutionary in themselves, can cause grandiose intellectual revolutions. This was the case with Copernicus, as it would later be with Darwin. These texts were read – if not precisely understood – by a growing number of non-specialists; they captured both the intel-lect and the imagination of their readers, did away with old and well-established answers, and created a host of new problems. For instance, the work of Copernicus provoked questions such as these: what is grav-ity and why do heavy bodies fall on the surface of a moving Earth? What moves the planets and how do they remain in orbit? How big is the universe and what is the distance between the Earth and the fixed stars? Questions were also being raised outside of scientific circles. To accept terrestrial motion and a new world system led not only to an upheaval in astronomy and physics and the need for their restructuring but also to changes in man's ideas about his world, about nature, and about his place in it. Every unstable system (and without a doubt this was the con-dition of astronomy in Copernicus's time) has its problematic points that cannot be touched lest the whole system collapse. The movement of the Earth was one of these.

The World in Pieces

In one of his 1539 *Table Talks*, Luther makes reference to a "fool astronomer" who claims that the Earth moves, intends to turn all of astronomy upside down, and challenges Scripture where it says that Joshua ordered the sun and not the Earth to stand still. Six years after the publication of Copernicus' masterwork, Philip Melanchthon emphasized in his own *Elements of Physics* that those who believed that the eight spheres and the sun did not revolve around the Earth held pernicious notions that "want of honesty and decency." And Calvin, without even mentioning Copernicus, strenuously reaffirmed the literal truth of Scripture.

There has been much discussion of both Protestant and Catholic reaction to Copernicanism. One of the best known historiographic legends maintains that the Roman Curia and Scholastic theologians remained basically indifferent to it. In 1546, only three years after the death of Copernicus, the Dominican Tolosani, an associate of Bartholomew Spina's and teacher at the Sacred Palace, voiced the Roman Curia's views on Copernicanism as quasi-official spokesperson, and criticized the new system in a document that came to light only in 1975, *De veritate Sacrae Scripturae*. According to Tolosani, Copernicanism had one essential flaw: it violated the fundamental and unrenounceable principle of *subalternatio scientiarum* for which "a lower science requires a higher science." This was no small consideration, for the first among sciences – theology – offered the cosmologist a description of the physical universe, and no science could conflict with theology. Copernicus, though an able mathematician and astronomer, was found to be "less able in the physical sciences and dialectic, and inexperienced in Scripture." Tolosani's paper was carefully studied by another Dominican, Tommaso Caccini, whose violent attack on Copernicanism on December 20, 1614, from the pulpit of Santa Maria Novella in Florence, was the basis of the Church's 1616 condemnation of Copernican theory as "philosophically foolish and absurd and expressly heretical." In his own prefatory letter of dedication to Pope Paul III, Copernicus had appealed to his influence and judgment in order to "hold the slanderers from biting, though the proverb hath it that there is no remedy against a sycophant's tooth" (Camporeale, 1977–8; Garin, 1975: 283–95; Kuhn, 1957: 143).

Indeed, there were to be many bites over time, but as so often in the face of what is new there was also cautious acceptance from experts, enthusiasm, disapproval, and above all expressions of doubt and confusion. *De*

Revolutionibus was reissued in Basel in 1556, thirteen years after its initial publication, and included Rheticus' *Narratio Prima* in the appendix to help explain the significance of the new world system to non-experts. Rheinhold's *Prutenic Tables* (1551) were revised and added in 1557. One year earlier, *The Castle of Knowledge* by physician and mathematician Robert Recorde (1510–ca. 1558) was published in London. In a dialogue between Master and Scholar, the former says that it is too soon to discuss whether the Earth moves because its stability is so fixed an idea that to suggest the opposite would be mad, and the Scholar observes that widely-held opinions are not always true.

For the most part though, astronomers proceeded with caution. With the notable exceptions of Kepler and Galileo, they rejected the declaration of a system that was superior to Ptolemy's. The success of the new tables led to a reaction typified by Thomas Blundeville's 1594 remark that Copernicus had, with the help of a false hypothesis, succeeded in making more exact demonstrations than had ever before been made. Michael Maestlin (1550–1631), professor of astronomy at Tübingen, added an appendix that explained the Copernican system to the final edition of his *Epitome of Astronomy* (1588). Since Maestlin had been Kepler's teacher we can assume that he instructed his pupil in the Copernican system. He also helped Kepler write and publish the *Mysterium cosmographicum* (1596) (which included difficult calculations), for which the author gave him a gold-plated silver goblet and six silver dollars in appreciation. In 1587 or so, Christoph Rothmann, astronomer to the Landgrave William IV of Hesse-Kassel, passionately defended Copernicanism in letters to Tycho Brahe. He refuted traditional arguments against the movement of Earth and claimed it was impossible to defend a literal interpretation of Scripture that would force one to believe in the existence of "water in the heavens," a question that had been of vital importance throughout the Middle Ages.

In his *Book of Diverse Speculation on Mathematics and Physics*, the mathematician Giovanni Battista Benedetti (1530–90) argued that Aristotelian ideas were of no value against Copernicus. Among philosophers, in addition to Thomas Digges and Giordano Bruno, there was also the point-of-view of Francesco Patrizi da Cherso (1529–97). A professor of Platonic philosophy, who taught first at the University of Ferrara and then was called to Rome by Clement VIII, Patrizi's cosmological views were a bizarre mix from today's perspective. In Patrizi's system, the Earth was still the center of the universe and the sun revolved around it, yet the Earth indeed moved according to Copernican theory, though only diurnally (Patrizi rejected the two other movements described by Copernicus). The stars were not fixed in actual spheres, but like large animals that move on their own, were powered by their own soul. There was a single, continuous and fluid sky. The fixed stars only appeared to move and depended on the diurnal rotation of the Earth on its own axis. The stars were not all the same distance from Earth, and were scattered about an infinite universe.

Astronomers may not have been pleased, but the dividing line between those who rejected Copernicanism and those who accepted it, or doubted the novelty of it all, was not simply one between the professional astronomers and philosophers or literary intellectuals. Those who first supported Copernican theory in England, for example, were neither radicals nor advocates of a new scientific method. Robert Recorde felt that astronomy was the handmaiden of astrology. The Copernican mathematician John Dee (1527–1608) authored both a celebrated "Preface to Euclid" and the *Monas hieroglyphica* (1564), a book about the secrets of super-celestial virtues as revealed by the mysteries of the Cabala, Pythagorean numerical combinations, and the Hermetic Seal. Thomas Digges (1543–75) made references to the occult (recalling Hermes Trismegistus and the 1534 poem *Zodiacus vitae* by Stellatus Palingenius) in an appendix he authored and added to the 1576 edition of his father's *Prognostication Everlasting*. Entitled *Perfit Description of Caelestiall Orbes,* he described an immovable orb of fixed stars in an infinite sphere with no upper boundary that is "the palace of felicity [. . .] the very court of celestial angels devoid of grief and replenished with [. . .] the habitacle for the elect." Giordano Bruno (1548–1600) was in London in 1585 and took to defending the Copernican world view. In *On the Infinite Universe and Worlds* (1584), Bruno presented Copernican theory against a background of astral magic and sunworship and linked it to the main themes in Ficino's *De vita coelitus comparanda.* He argued that the Copernican "diagram" revealed the "hieroglyphics" of God: the Earth moves because it lives around the sun; the planets, like living stars, journey alongside her; other innumerable worlds, which live and move like great animals, inhabit an infinite universe. And the writings of William Gilbert, himself something of a "Copernican," also included animistic themes and references to Hermes, Zoroaster, and Orpheus.

Because heliocentrism was often associated with magic and the occult, followers of Copernican ideas were often drawn into a larger battle against neo-Platonism. Francis Bacon's views on Copernicanism (from the period 1610–23) should be considered in light of such great uncertainty and ambiguity, for all too often they have been viewed (by spiritualists in the late nineteenth century and neopositivists and Popperians in the twentieth century) as an expression of an ahistorical condemnation. It is hardly fair to consider a period so plagued by uncertainty and doubt as "scientifically backward." Bacon had been very excited by Galileo's discoveries of 1612, but himself died in 1626. Mersenne's "conversion" to Copernicanism took place between 1630 and 1634. In his 1634 *Novarum observationum libri*, the mathematician Gilles Personne de Roberval (1602–75) declared that he was unable to choose which of the three competing world systems was the correct one because it was entirely possible for "all three of the systems to be false and the true one yet undiscovered."

The bylaws of the University of Salamanca in 1561 dictated that the study of mathematics would include Euclid and either Ptolemy or Copernicus, the choice of which was up to the students. They almost never chose Copernicus, but this made Salamanca a true exception. Up until the seventeenth century, most universities (even those in Protestant countries) taught the two (or three) world systems side-by-side. It is also important to keep in mind that in the first decade of the seventeenth century those who rejected the existence of the celestial spheres, like Brahe, Gilbert and Rothmann, were not academics. Their number grew substantially only in the 1620s, and the doctrine was finally eliminated from astronomy texts in the 1630s. Cultural acceptance of the new world system required the answering of thorny questions that were not just astronomical in nature. The greatness of men like Galileo and Kepler lies in part with their choice of Copernicanism. Both men looked to Copernicus as their master and contributed to the astronomical revolution begun by him, though their work did not effortlessly open up new venues. These lines from the poem *Anatomy of the World* (1611) by John Donne (1573–1631) capture the loss felt by so many as their familiar and reassuring world collapsed before them:

> And New Philosophy calls all in doubt,
> The Element of fire is quite put out;
> The Sun is lost, and th'Earth, and no man's wit
> Can well direct him where to look for it.
> And freely men confess that this world's spent,
> When in the Planets, and the Firmament
> They seek so many new; then see that this
> Is crumbled out again to his Atomies.
> 'Tis all in pieces, all coherence gone;
> All just supply, and all Relation:
> Prince, Subject, Father, Son, are things forgot,
> For every man alone thinks he hath got
> To be a Phoenix . . .

Tycho Brahe

Earlier in this chapter, a *third* world system was mentioned. Tycho Brahe (1546–1601), self-taught Danish astronomer who studied in Leipzig (with-

out formally matriculating at the university), was very interested in alchemy and also firmly believed in an affinity between celestial events and
terrestrial phenomena. The illustration on the title-page of his *Astronomiae
instaurate mechanica* shows Brahe sitting on a globe with a compass in
one hand and his eyes cast to the heavens. It is labeled: *suspiciendo despicio*
(looking up, I look down). In another illustration a snake (symbol of
Aesculapius) is wrapped around his arm, and he is looking at chemistry
equipment. It is labeled: *despiciendo suspicio* (looking down, I look up).

   More than a natural philosopher, Tycho was a patient and meticulous
observer, perhaps the greatest naked-eye observer in the history of astronomy. He made his first observations in 1563 at the age of sixteen,
and his life-long progressively more precise observations have stunned
many the historian of astronomy. He used instruments both crafted by
others and sophisticated ones of his own manufacture. Unlike many of
his professional colleagues, he observed the planets continuously and not
just when they were in a favorable configuration.

   While on his way home on the night of November 11, 1572, Tycho
(then twenty-six years old) saw a bright, new star in the constellation
Cassiopeia. This was a turning point in his life. Brahe decided against an
opportunity to move to Basel, and his observations brought him to the
attention of the King of Denmark who gave him a feudal title to the
island of Hveen. There he built the splendid castle of Uraniborg, full of
observatories and laboratories, which became the training ground for
many young European astronomers. As bright as Venus at its brightest,
the new star grew progressively dimmer until it disappeared entirely in
early 1574. Kepler later wrote that the star, "if it produced nothing else,
produced a great astronomer." Tycho recorded his observations in *On
the New Star* (1573). If this new star was not a comet, and if it appeared
in the same position with respect to the sphere of the fixed stars, then he
had witnessed a change in the immutable heavens, and so it became necessary to question the difference between the unchanging heavens and
the corruptible sublunar world. Tycho was certain that his observations
of the comets of 1577 and 1585 confirmed his theory. He tried to measure the parallax of the 1577 comet: it was too small to belong to sublunar
regions. He wrote that "[a]ll comets observed by me moved in the aetherial
regions of the world and never in the air below the moon, as Aristotle
and his followers tried without reason to make us believe for so many
centuries." If then the comets were located *above the moon*, the planets
could not be fixed in the traditional crystalline sphere. In a letter to Kepler,
he wrote that he believed that "the physical reality of the spheres must be
banished from the heavens." Comets follow the laws of no sphere, but
act "in contradiction to those laws." The machine of Heaven is not a
"hard and impervious body full of various real spheres, as up to now has

been believed by most people [. . .] It [. . .] extends everywhere, most fluid and simple, and nowhere presents obstacles as was formerly held, the circuits of the Planets being wholly free and without the labor and whirling round of any real spheres, being divinely governed under a given law." The spheres "do not actually exist" in the heavens and "are allowed only in the interest of learning" (Kepler, 1858–71: I, 44, 159).

This statement was as revolutionary as Copernicus's declaration that the Earth moved. One of the central tenets of traditional cosmology – the incorruptibility and immutability of the heavens – had collapsed, right there on the field of astronomy (and not, as in Patrizi's case, in the arena of speculative imagination). In chapter 8 of *On the Most Recent Phenomena of the Aetherial World* (the very use of the word *recent* was a challenge to convention), published at Uraniborg in 1588, Tycho outlined the world system that he created out of necessity after rejecting both the Ptolemaic and Copernican systems. While the Copernican system was elegant and mathematically superior to Ptolemy's, Tycho did not accept the notion that the "large and sluggish Earth" was capable of movement (not to mention three of them). If the Earth indeed moved, he observed, then a stone dropped from a tower would not land at the foot of the tower (as it does in reality). The Copernican system was moreover unacceptable because the distance between Saturn and the fixed stars would be enormous given the absence of an observable parallax of the stars. Finally, the Copernican universe contradicted Scriptural passages which refer to the immobility of the Earth. A new system would have to "agree with Mathematics and Physics, avoid theological censure, and at the same time wholly accord with the celestial appearances" (Brahe, 1913–29: IV, 155–7).

In Tycho's system the Earth is fixed at the center of a universe bounded by a sphere of stars that rotates daily and accounts for their daily movement. As in the Ptolemaic system, the sun and moon both revolve around the Earth. The five other planets (Mercury, Venus, Mars, Jupiter, and Saturn) revolve around the sun. Tycho rejected the solid nature of the spheres because the orbits intersected in several points. For his system to work, it was necessary to retain epicycles, eccentrics, and equants.

The Tychonic system was mathematically equivalent to the Copernican system. It neither contradicted Scripture nor made it necessary to reject such a deeply-rooted principle as the immobility and centrality of the planet Earth. The system was attractive to those who rejected the Copernican revolution, and preferred by many Jesuits. Though Tycho's reputation as a great astronomer undoubtedly hindered the spread of Copernican theory, the questions his work raised contributed to the downfall and gradual abandonment of the Ptolemaic system of the world.

Kepler

Johannes Kepler (1571–1630) was born to Lutheran parents in Weil, a small town in Württemberg. With the intention of becoming a minister, he attended the Protestant University of Tübingen where Maestlin instructed his students in both Ptolemaic and Copernican astronomy. Kepler became the district mathematician of Styria and also taught mathematics at the Protestant seminary in Graz, Austria. Among his varied duties was the preparation of "forecasts," one of which foretold a combination of cold winter, peasant revolt, and war with the Turks. Nor did he cease casting horoscopes later in his career, some of which – Wallenstein's for example – were insightful psychological profiles. In 1595 he wrote the *The Cosmographical Mystery*, and with the help of Maestlin, published it in 1596.

Kepler's scholarship has always been considered unusual by historians. Unlike most scientific writers, he did not limit himself to explaining the results of his studies. He also discussed the underlying reasons for his theories, and described both the trials and errors of his method. He felt that an explanation of what drove him to write a book was essential to its comprehension. He tells us that, already dissatisfied with the inadequacy of the traditional system, he was overcome by enthusiasm when he heard the Copernican system described to the point of defending it and beginning a study of not just the mathematical reasons for the movement of the sun (which Copernicus had done) but also the "physical and metaphysical reasons." In Kepler's opinion, the Copernican system respected celestial phenomena and could explain past movements and predict future ones with greater precision than Ptolemy or other astronomers. Ancient philosophers constantly needed to invent spheres, while Copernicus had simplified the cosmological machine; this greater simplicity belied the essence of the system itself because nature loves simplicity and unity never harbors anything idle or superfluous.

Yet the *Cosmographical Mystery* was not primarily intended as a defense of Copernicus. It was a demonstration that God, in creating the world and the heavens, had "looked at the five regular solids that have been so famous since the time of Pythagoras and Plato" and created harmony between their nature and the number, proportion, and relationship of the celestial motions. The five regular or "cosmic" solids to which Kepler referred had one special feature in common: they are the only ones whose faces are all identical and equilateral figures: the cube, tetrahedron, dodecahedron, icosahedron, and octahedron. Kepler asked then what were the *causes* behind the number, size, and motion of the planets and said that his study was motivated by the incredible correspondence between, on the one hand, sun, fixed stars, and the intermediate space and, on the other, the Holy Trinity. But his investigation into the possibility that one planetary orbit was twice, three times or four times greater than another yielded little, nor did the introduction of planets so small they were undetectable between one orbit and another. The five regular solids appeared, after many unfortunate trials, to be a way out and so seemed to Kepler to be an extraordinary discovery. The celestial orbits, which according to Copernicus numbered six, corresponded to just five figures that "from the infinite number that is possible have unusual properties that no others have." The orbit of the Earth became the measure of all the others. If the sphere of Saturn were circumscribed about the cube in which Jupiter's sphere was inscribed, and if the tetrahedron were inscribed inside Jupiter's sphere with Mars' sphere inscribed in it, and so on (in the order given above for the figures), then the relative dimensions of all the spheres would be those calculated by Copernicus. There were actually some discrepancies, but Kepler hoped that more accurate calculations could correct these, and so he turned to the work of Tycho Brahe.

In the *Cosmographical Mystery*, Kepler tried not only to discover the structural laws of the universe but also to answer the questions of *why* planets moved and *how fast* (much slower the farther they are from the sun). He believed it was necessary to accept one of the following two statements: either the motive force of the individual planets is weaker the greater its distance from the sun, or only one motive force exists – the sun – that drives all bodies: with greater force for nearby bodies and weaker for distant ones because force diminishes with distance. Kepler chose the second and claimed that force was proportional to the circle in which it spreads and decreases with distance. Given that the period increases with a widening circumference "a greater distance from the sun acts doubly in increasing the period, and inversely, half of the increase in period is proportional to the increase in distance." His calculations were not too far off from those of Copernicus and he felt he had "gotten close to the truth." The sun lay at the center of Kepler's universe (whereas the center of the Copernican universe was not the sun but the center of the Earth's orbit); it was the center of all life and motion, and was the soul of the universe. The fixed stars were at rest and planets moved as a result of a secondary or external motion. The principal movement or force was that of the sun, the most splendid and beautiful of all creations, and it was more noble than all secondary motion. Fixed and the source of all movement, the sun was the very image of God the Father. Not just the universe but all of astronomy became heliocentric. The sun was not just the architectural center of the cosmos but its *dynamic center* as well.

Maestlin was impressed by *The Cosmographical Mystery*, and the young Kepler sent it to Tycho Brahe. Galileo saw a copy of the book and wrote to Kepler, congratulating him on his Copernican views. However, he most likely had not yet read it. When Kepler later attempted to establish a correspondence with him, he received no response. Indeed, Galileo's distance from all forms of mysticism estranged him from the type of science practiced by Kepler. It also proved to be an obstacle to Galileo's appreciation of all of Kepler's important discoveries in the field. On the other hand, Kepler's contact with Tycho Brahe, who was more receptive to Hermetic and mystical views, was of great importance.

Tycho wrote to Kepler that the harmonies and proportions of the universe had to be discovered *a posteriori* and not determined *a priori*. Aside from this basic reservation, Tycho thought highly of the *Mystery*. He had by now left Denmark and was Imperial Mathematician of Bohemia, and so offered Kepler a job as his assistant. In 1600 Kepler began developing a new theory of the orbit of Mars so he could prepare new tables to replace the *Prutenic Tables*. The *Rudolphine Tables* were in fact not published until 1627. Brahe's death in 1601 changed Kepler's situation, and he succeeded Tycho as Imperial Mathematician, gaining access to all his papers.

It was during this period that Kepler wrote the following books in addition to his work on almanacs and forecasts: *De fundamentis astrologiae certioribus* (1601); *Ad Vitelionem paralipomena* (a fundamental work in the history of optics, 1604); *De stella nova* (1606); *De Jesu Christi Salvatoris nostri vero anno natalitio* (1606). In 1606 he also completed his masterpiece, *Astronomia nova seu Physica coelestis*, though it was only published in 1609, the same year that Galileo began observing the heavens with a telescope.

In *Astronomia nova*, Kepler described in detail all 60 of his attempts to reconcile Tycho's data for the orbit of Mars with the various combinations of orbits given by Ptolemaic and Copernican astronomies. The difference between his predictions and Tycho's observations was only eight minutes of arc. Though this solution may have satisfied the astronomers of the period, Kepler abandoned the problem, and despairing of finding an acceptable solution, decided to work on the orbit of the Earth. It moved fastest when it was closest to the sun and slowest when farthest from it. Though this work was based on an erroneous assumption (that the Earth's velocity is inversely proportional to its distance from the sun), and he made some gross mathematical errors, Kepler nevertheless managed to deduce what we today call *Kepler's second law*: a line joining a planet to the sun sweeps through equal areas in equal times. What Kepler's solution implied – and the old astronomy and Copernicus himself did not – was that the Earth and the other planets followed a *real* and not just seemingly non-uniform path.

A simple geometric law could explain this non-uniformity. The physical causes of the variation lay once again with the sun. Kepler looked not only to Tycho and Copernicus as his teachers, but Gilbert as well. Gilbert's magnetic philosophy gave Kepler what he needed to explain the physical variations of velocity. Kepler explicitly referred to the soul inherent in heavenly bodies. Yet unlike Patrizi or Bruno, he not only made mathematical calculations and precise astronomical observations but also questioned *how* these souls or motive spirits worked. The basic propositions of Aristotelian physics were at the heart of his way of thinking and his move to join celestial to terrestrial physics. For Kepler, in this regard an Aristotelian, only the application of a force could explain continuous motion. Kepler knew neither the principle of inertia nor the notion of centripetal force. The power that radiated from the sun did not exert a central attraction: it served to move the planets and *keep them moving*. In *Astronomia Nova*, even when Kepler rejected the explanation of a specific soul for each planet, his assignment of a single soul to the sun was in no way a sort of "concession" to animistic metaphysics. The very motors of the planets were attractions like "the attractions in a magnet that tends toward a pole and attracts iron." All of planetary movement was therefore governed "by purely physical, that is magnetic, powers." There remained however one exception that was necessary for the operation of the system: a motive soul was responsible for the rotation of the sun on its own axis. Kepler did not attribute rotation to the moon. But the sun, the central body of the universe, *must* rotate on its axis and pull along the entire body of the universe: "the sun rotates on its own axis like a beacon atop a tower and emits an immaterial *species* of its body similar to the immaterial *species* of its light. This *species*, as a result of the sun's rotation, rotates in the form of a high-speed vortex that spreads throughout the vast universe and carries the planets along with it."

Kepler broke with a time-honored tradition when he declared that the orbit of a planet was not a circle but "passes gradually inside at the sides, and again increases to the amplitude of a circle at perigee. The shape of this kind of path is called an oval." The shift from oval to ellipse proved extremely complex, and Kepler describes in great detail the computational errors he makes and difficulties he encounters. Only a perfect ellipse with the sun at one focus (an idea that came to him like a proverbial bolt of lightning) fit the observational data and the law of areas; that realization came to be known as Kepler's First Law. One conic curve was sufficient to describe the orbit of each planet, and by abandoning the dogma of circularity, there was no longer any need for eccentrics and epicycles, resulting in a greatly simplified system. As Richard Westfall has pointed out, "if it is true to say that Kepler perfected Copernican astronomy it is equally true to say that he destroyed it" (Westfall, 1971: 12).

Kepler had used complex mathematical language to explain heavenly

phenomena to a select audience in *Astronomia Nova*. He now planned
to write a book in question and answer format that would sum up the
new astronomy and serve as a manual (to replace the one currently in
use). In 1610 he published *Dissertatio cum Nuncio Sidereo* and in 1611,
the *Dioptrice*. In 1612, after the abdication of Rudolph II, Kepler left
Prague and moved to Linz where he stayed for 14 years. War forced
him to give up his job as mathematician for the Austrian city. Despite
his wishes, he never returned to Germany. He worked for a number of
patrons (including Wallenstein) and died in Regensburg in 1630.

The various books of Kepler's summary/manual, *Epitome of Copernican Astronomy*, were published between 1617 and 1621. His astronomical discoveries were discussed in the context of the Pythagorean and
Neoplatonic ideas that characterized his earlier work, the *Mystery*. Light,
heat, movement, and the harmonies of motion were the perfection of the
world and soul-like entities. The sphere of the fixed stars "retains the
heat of the sun so it will not be lost and acts like a wall or skin or suit for
the universe." By virtue of its body, the sun was the *cause* of planetary
motion. The ether found a series of terrestrial corrolaries: its vegetable
corresponded to the nourishment of animals and plants; its heat to the
force of life; its movement to animal force; its light to sensory force; and
its harmony to reason. Movement could not be explained solely as an
impetus conferred by God upon the sun at the moment of Creation: "its
constancy and perpetuity, upon which all life in the world is based, is
most easily explained as the expression of its soul."

Such "Pythagorean" themes were all the more evident in the *Five Books
of the Harmonies of the World*, published in Linz in 1619. Like the
Epitome, this work had long been in the planning stages; in 1600 Kepler
had written to Herwart of Hohenburg, "would that God deliver me from
astronomy so I can devote all my time to my work on harmonies." Alongside the geometric relationships employed in the *Mystery* (to which Kepler
would add a sixth figure, the star-shaped polyhedron), Kepler considered harmonic relationships – given that God was not just a geometer
but also a musician. The table of contents of the fifth book showed that
every planet corresponded to an individual musical note or interval. Planetary counterpoints, or universal harmonies, differ from one another,
and the planets express four types of voices: soprano, contralto, tenor,
and bass. The third chapter of the fifth book not only summarized the
central idea of the *Mystery* but also added a new theory: It is an absolutely certain and exact fact that the ratio between the periodic times of
any two Planets is precisely as the three-halves power of that of their
mean distances, that is, of the orbits themselves. This was Kepler's Third
Law: the squares of the periodic times of any two planets are proportional to the cubes of their respective mean distances from the sun. Once

the orbit has been established so has the speed and vice versa. And so a law had been discovered that was not limited to regulating the planetary motion of individual orbits: it established a relationship between the speeds of planets moving in different orbits. In Kepler's eyes, the discovery of the third law was like a great metaphysical discovery: "*Gratias ego tibi, Creator Domine.*" For Kepler the book might be read in his own days or by a subsequent generation. He imagined it might even wait a hundred years to find a reader; after all, "had not God waited six thousand years for someone to contemplate his wonderful works?"

Kepler's path of discovery was long and tortuous, and only Alexandre Koyré (Koyré, 1961) has had the perseverance to reconstruct it in an analytical way: not only did Kepler derive the second law from "erroneous" assumptions but he established it as truth *before* determining the elliptical nature of planetary orbits. The three laws that bear Kepler's name emerged in an atmosphere – taking Descartes and Galileo as points of reference – that can hardly be considered "modern."

Historians have always emphasized the extraordinary blend of number mysticism and passion for observation present in the work of Kepler. Many have dwelt on his determination to search for data that would fit imaginative metaphysical hypotheses. Many have associated Neo-Pythagorism and Hermeticism with Kepler, to the point of identifying him with these traditions. Kepler has certainly been a cumbersome presence in the spectrum between Galileo and Newton. For all his mystic leanings, unlike Patrizi and the magicians and natural philosophers of the late Renaissance, Kepler was deeply interested in *how* the souls of heavenly bodies functioned. Beyond an abiding belief in the mystic potential of Neoplatonism, Kepler's "modernity" can be linked to two themes: (1) an inquiry into the quantitative variations of the mysterious forces that operated in time and space; (2) the partial abandonment of an animistic point-of-view in favor of a more mechanical one. Both motion that occurs in space and the "virtue" radiated by the sun through space are "geometrical things." From this perspective, the celestial machine "should not be compared to a divine organism but rather to the mechanism of a clock." Its movements occur "thanks to a very simple, single magnetic force like the simple weight that causes the movements of a clock."

The idea that the world was *not* a divine organism placed Kepler squarely against the late Renaissance occult tradition. Kepler himself believed that the reduction of many spirits (of the individual planets) to a single spirit (of the sun), and the identification of the soul with a force were positive outcomes. In an annotated edition of the *Mystery* (1625), Kepler stated that in *Astronomia Nova* he had proved that specific spirits for the individual planets did not exist, and claimed that with regard to the sun, "if you substitute the word *force* for the word *soul*, you have the very principle on which my celestial physics is based." Once, he wrote, "I firmly believed that a soul or *anima* was cause for planetary movement." Reflecting on the fact that the mover grows weaker in proportion to distance and the same is true for the light of the sun, "I came to the conclusion that this force was corporeal, even if I do not mean corporeal in the literal sense but as a metaphor, in the way we say that *light* is something corporeal."

Kepler's mysticism was tied to a specific belief: that truth could not be discerned in symbols and hieroglyphics but through mathematical proofs, without which, he wrote to the magician Robert Fludd, "I would be blind." It was not a case of "finding pleasure in things cloaked in mystery" (as was the case with magic) "but of clarifying them." The first method "is that of alchemists, Hermetics, and Paracelsians while the second is the exclusive property of mathematicians."

It surely must have been difficult at the time to understand these differences, to embrace scientific findings masked as divine revelation, and to work within a system of ideas that held neither the all-too-familiar problems of ancient astronomy nor the clear ideas of the new philosophy. Galileo not only emphasized the profound difference between Kepler's "philosophizing" and his own, but maintained that some of Kepler's ideas were "more detrimental to Copernican doctrine than supportive (Galileo, 1890–1909: XVI, 162; XIV, 340). Bacon, who was in many ways tied to the Hermetic tradition, ignored Kepler altogether. In a letter to Mersenne on March 31, 1638, Descartes called him "my first teacher of optics" but did not consider the rest of his work worthy of mention. Only Alphonse Borelli (1608–79) understood the significance of Kepler's astronomy. Kepler's laws were finally considered "scientific" only after Newton found a use for them, and they were accepted by the majority of astronomers only in the 1660s.

6
Galileo

Galileo Galilei was born in Pisa, Italy on February 15, 1564. His Florentine father, Vincenzio Galilei, was a merchant and an accomplished musician and music theorist. His mother, Giulia Ammannati, was from neighboring Pescia. Though Galileo enrolled in the University of Pisa in 1581 with the intention of studying medicine, he studied mathematics instead. In 1585 he left the university before completing his degree. The first fruit of his interest in physics and the Archimedean method was *Theoremata circa centrum gravitatis solidorum*. Basing his work on Archimedes, he designed a hydrostatic balance in 1586 and wrote *La bilancetta*.

Galileo's first job as professor of mathematics at the University of Pisa came in 1589 thanks to the influence of Guidobaldo del Monte with the Grand Duke Ferdinand. It was during his tenure at Pisa that Galileo wrote *De Motu* (ca. 1592), a work in which he made the anti-Aristotelian claim that all bodies are intrinsically heavy and lightness is simply relative: for instance, flames rise not because they possess the characteristic of lightness but because they are lighter than air. In this same work Galileo discussed the velocity of different bodies in a like medium, as well as like bodies in different mediums and different bodies in different mediums. He was not trying to prove that all bodies fall at the same speed, but that the speed of the fall of a body is proportional to the difference between its own weight and the density of the medium through which it falls. Like objects that are equally dense will fall through the air at the same speed, regardless of their weight. However, if two objects of equal weight but different composition were dropped at the same time, the denser object will fall faster. So, contrary to Aristotle, given the progressive reduction in the density of the medium, motion in a vacuum is indeed possible and objects made of different materials fall through it at different speeds.

This work marked the beginning of a long journey which ultimately led Galileo to reject Aristotelianism. He worked on a variety of problems over the next fifty years: the isochronism of a swinging pendulum; falling bodies; projectile motion; cohesion; the strength of materials; "impulsive" or impetus force. His theories and methods changed many times over the course of this half century as he incorporated corrections, a deeper understanding of the problems, and even conceptual developments into his work. One element which remained constant, however, was a tacit acceptance of both the approach and method of the "divine Archimedes."

Galileo's interest in technical problems, first expressed in *La bilancetta*, continued even after he became a professor of mathematics at the University of Padua in 1592. Between 1592 and 1593 he wrote three works – *A Brief instruction in military architecture*, *On Fortifications*, and *On Mechanics* – which were not even published until 1634, and then in French by Mersenne. He lectured on Euclid's *Elements* and Ptolemy's *Almagest*. In 1597, he wrote *Treatise on the Sphere, or Cosmography*, a clearly worded guide to the geocentric system for students. Yet he was already following a different path. In a letter to Kepler that same year, he revealed that he had long been a convert to Copernicanism even if he had not yet dared publish his findings out of fear of what had befallen their common master. Nor did he abandon his other interest. The instruments he needed for his lessons and other pursuits were all crafted in a small workshop next to his study. From the workshop came objects related to his work on military architecture and fortifications, ballistics, hydraulic engineering, in addition to instruments for his research on the strength of materials, the construction of the geometrical and military compass, the telescope, and thermo-baroscope. His passion for observation, measurement and instruments never abated, nor did his infinite curiosity for experiments. He published a pamphlet about the *Geometrical and Military Compass* in 1606, and the next year wrote the *Defense against the deception and fraud of Baldessar Capra*, who had falsely claimed to be the inventor of the compass.

Astronomical Discoveries

The year 1609 was important in the history of science. The great astronomical discoveries of the year (the *Sidereus nuncius* appeared in 1610) not only undermined a consolidated view of the world but also weakened the claims against the Copernican system. To begin with, the moon was terrestrial in nature and nonetheless *moved* through the sky: from this perspective, it was no longer ludicrous to suggest that the Earth, too, moved. What was more, Jupiter and its revolving satellites suddenly appeared to be a small-scale version of the entire Copernican system. Observation of the fixed stars revealed that they were much farther away from the Earth than the planets were, and they did not appear to be located directly behind the sky of Saturn. The absence of an observable parallax of the stars had been one of the biggest objections to the Copernican system. Parallax is the change in position that occurs when the same object is observed from different places (if you look at a pencil with one eye closed and then open that eye and close the other, the pencil looks like it has moved). The greater the distance the smaller the change should be. The objection advanced by Tycho, among others, was this: if the Earth moved through space, the shape of the constellations should change from season to season. The impossibility of detecting stellar parallax could now be explained by the vast distance that separated the stars and the Earth.

Fresh evidence in support of the Copernican system and against the Ptolemaic came in the form of the astronomical observations made by Galileo shortly before leaving the University of Padua for a court appointment in Florence as "Chief Philosopher and Mathematician to the Grand Duke" in September 1611. He had observed the "tricorporeal" appearance of Saturn (the "rings" were not visible through his telescope), sunspots, and the phases of the planet Venus. The observation that Venus "showed phases just like the moon" rightfully struck Galileo as significant. Indeed this revealed a situation which in no way fit a Ptolemaic scheme of the world nor could be explained by it.

In a letter to Cesi in May 1612, Galileo announced that the "novelty" of the sunspots he had discovered was the "death or rather last judgment of pseudo-philosophy." The changing shape and disappearance of the spots on the very surface of the sun, he later explained in the *History and Demonstration Concerning Sunspots and their Phenomena*, were not an obstacle to "freethinkers" who had never believed that the heavens beyond the sublunar sphere were immutable (Galilei, 1890–1909: V, 129).

Following the great astronomical discoveries of 1610, Galileo threw all caution to the wind. He wrote to Giuliano de Medici in January 1611 that "we have sensory experiences and necessary demonstrations for two problems that until now have remained unresolved by the greatest thinkers of the world" (ibid.: XI, 12). One was that planets were opaque bodies, and the other that they rotated around the sun. While the Pythagoreans, Copernicus, Kepler, and Galileo himself had "believed" this to be true, it had never been "reasonably proved." Kepler and other Copernicans could now glory in the knowledge of "having believed and philosophized well, even though we have been and still are considered by all *in libris* philosophers to not be very knowledgeable and little more than fools" (ibid.: XI, 12).

In hopes of securing the court appointment he so desired – chief philosopher in addition to chief mathematician – Galileo wrote to the grand duke's secretary of state describing his future projects almost immediately after the *Starry Messenger* was published. He said that he was planning two books on the system and the constitution of the universe; three books on local motion ("an entirely new science in which no one else, ancient or modern, has discovered any of the remarkable laws which I demonstrate"); three books on mechanics; and even works on sound, the ocean tides, the nature of continuous quantities, and the motion of animals. The new astronomy and physics would not only prove Copernicanism but also establish a new science of nature. To the "viper-like obstinacy" of academic philosophers and professors, Galileo now proudly proposed his own philosophy and declared that he had "studied more years of philosophy than months of pure mathematics" (ibid.: X, 353).

Galileo's remarkable confidence owed something to the fact that he was indeed appointed Chief Philosopher and Mathematician to the Grand Duke of Tuscany in September 1611 (and so moved from Padua to Florence). In light of recently discovered documents, his decision to leave Padua was actually rife with consequences. Until the 1992 discovery by Antonino Poppi of an earlier document, it has long been accepted that the Rome Holy Office first expressed concern about Galileo and his work during the congregation of May 17, 1611, when it was explicitly asked if Galileo had been mentioned in the proceedings against Cesare Cremonini. The Poppi document instead shows that seven years earlier, on April 21, 1604, "he had been formally accused of heresy and libertine customs by the Inquisition of Padua." Although the informer (most likely Silvestro Pagnoni, Galileo's scribe) admitted that "in matters of faith I had never heard him say anything negative," he accused him of casting horoscopes, failing to attend mass and receive the sacrament, having a lover, and reading unedifying books. "I learned from his mother," states the informer, "that he never went to confession or took holy communion, and she sometimes made me check to see if he attended mass on feast days; but instead of going to mass he would visit the Venetian whore Marina, who lives at Pontecorvo." (The woman to whom he referred is Marina Gamba, who bore Galileo three children between 1601 and 1606: Virginia, Livia, and Vincenzo.) He added, "I believe that his mother has gone to the Florence Holy Office against her son's wishes, and that he called her unspeakably rude names like whore and *gabrina*." If this last statement is indeed true, then Galileo was first denounced to the Inquisition as early as 1592.

In light of this new information, it would certainly seem that Galileo's move from Padua to the Tuscan court was not such a wise one. The Paduan professors who ultimately came under the suspicion of the Inquisition were strenuously defended by the Venetian Republic. The defense argued that "these accusations have been made by ill-meaning and self-interested souls [. . .] For this sound reason, and the knowledge that the reputation of the University will suffer, and divisions and disputes will arise among the students, we are moved to recommend that in Your customary prudence and sagacity You proceed no further in the investigation of these claims."

Admittedly, *if only* is a not a good phrase to use in the writing of history, but the evidence we have today surely sheds new light on Cremonini's words: "Oh, it would have been so much better if only Signor Galileo had not gotten involved in all this intrigue and had never left his Paduan freedom" (Poppi, 1992: 11, 58–60, 26–7, 62–3).

Galileo's sense of security was at any rate also tied to the events that followed his move to Florence. He was triumphantly received by scholars and clergy alike during his visit to Rome in 1611: he was elected to the *Accademia dei Lincei*, and was approvingly received by respected cardinals, the Jesuits, and Pope Paul V himself. In December 1612, Galileo felt nothing but hope and optimism. Yet storm clouds were gathering at that very moment. Galileo wrote a series of "letters" that were meant to persuade and convince a public of the new truths. However, in his optimism Galileo failed to detect the great cultural and "political" implications of the controversy over Copernicanism. He seemed sure, at the time, that victory was close at hand and believed he faced the ignorance and self-conceit of only individuals. He was unaware of both growing opposition in certain ecclesiastical circles and the general implications of his own position. He alternated between extreme confidence and an unchecked tendency to be controversial, rhetorical, and quick to find fault.

Nature and the Bible

It was not as if Galileo had not received warnings and advice to be cautious. Paolo Gualdo wrote him to "think twice before you publish this your opinion as fact, because many things can be said for the sake of argument that it would not be wise to assert as true." From the pulpit of St. Mark's monastery in Florence on the Day of the Dead in 1612, the Dominican Nicholas Lorini preached a sermon accusing Copernicans of heresy. At the end of the following year, Galileo's friend and faithful disciple Benedetto Castelli defended the doctrine of the Earth's mobility before the Grand Duke and his mother, the Grand Duchess Christina.

The growing controversy, along with a fear of losing favor with the Medicis, persuaded Galileo to direct action. His 21 December 1613 *Letter to Castelli*, widely circulated, explicitly addressed the question of the compatibility of Scriptural and scientific truth.

The text of the *History and Demonstration Concerning Sunspots and their Phenomena*, published by Prince Cesi that same year, was significantly censored. Galileo had written that the theory of the incorruptibility of the heavens was not only false but "erroneous and repugnant to the unquestionable truths of the Sacred Word which tells us that heavens and the whole world [. . .] were generated, dissolvable, and transitory." Cesi had to tell Galileo that the church's censors "having approved all the rest, want no part of this" (Galilei, 1890–1909: V, 238; XI, 428–9). The final approved version of the *Letter on Sunspots* contained no references to the Bible.

Scriptural decrees, wrote Galileo in the letter, are absolute and inviolable truths. Scripture is never wrong, though those who interpret it can make mistakes, especially when statements have been adapted for the easier comprehension of the Hebrews. As for the "naked sense of the word," many statements therefore appear "different from the truth," and have been adapted for the comprehension of ordinary people and require explanation by the wise interpreters. Nature and Scripture both proceed from the Word of the Lord: the first as the dictate of the Holy Spirit" and the second as "the faithful executrix of God's commands." However, while the language of Scripture has been adapted so that men may understand it and so its words may have different meanings, nature is instead "inexorable and immutable" and does not care if its causes and functions are "clearly grasped by man's intellect." In discussions about nature, Scripture "should be reserved to the last place." Nature has a logic and discipline not found in Scripture: "the Bible is not chained in every expression to conditions as strict as those which govern all physical effects." The "physical effects" that sensory experience sets before us can never be "revoked upon the testimony of Biblical passages which may have some different meanings." The task of the "wise interpreter of the divine text" (given that nature and Scripture can never contradict one another) is to "work hard to find the true sense of sacred passages" that agree with demonstrated scientific results. Moreover, given that Scripture admittedly is metaphorical, and that we are not in fact certain all its interpreters were divinely inspired, it would be prudent to allow no one to use Scriptural passages as proof of physical phenomena which could, in the future, be shown to be false. Scripture tends to persuade men of truths necessary for their own salvation, but it is not necessary to believe that what we know to be true through the senses and the mind has been given to us by Scripture. The second (and much briefer) part of the letter argues that the passage from Joshua (Joshua 10: 12) in which God wills him to stop the sun in order to make the day longer concurs perfectly with a Copernican universe and not an Aristotelian–Ptolemaic one (ibid.: V, 281–8; Galilei, 1957: 181, 182, 183, 212).

Galileo's artful attempt to divide his opponents by claiming that Copernican doctrine was compatible with the Bible still did not eliminate some difficult questions. For instance, what was the sense in stating that the passage in Joshua "clearly shows us that the Aristotelian–Ptolemaic universe is not feasible" if the Bible is a text concerned solely with salvation of the soul? And, once one compares the precise language of nature to the metaphorical language of the Bible, do not natural philosophers instantly become reputable interpreters of that language? What obligation do those who read and interpret the book of nature – also written by God – have to the interpreters of the Bible when Scripture concurs with physical reality? Finally, do they not inevitably trespass that ground reserved for theologians?

Many now felt that the union of theology and natural philosophy, which for centuries had seemed to guarantee the Church's authority in matters of conscience and intellect, was forever broken. Although Lorini presented a crude and approximate version of Copernican and Galilean theory in the charges of February 7, 1615, he was clear on a few points: in the *Letter to Castelli*, which "passed through the hands of everyone," Galileo had stated that in physical matters "Scripture should be reserved to the last place;" that the interpreters of the Bible often make mistakes; that the Bible "should not concern itself in matters other than those of faith," that in the case of physical events, "scientific explanations carry greater weight than sacred or divine ones" (ibid.: XIX, 297–8). Even Cardinal Bellarmine in 1615 insisted that the conclusions reached by the Council of Trent prohibited one from expounding the Bible "contrary to the common agreement of the holy Fathers." The Church Fathers and modern writers on Genesis, Psalms, Ecclesiastes, and Joshua "all agree in expounding *ad literam* that the Sun is in the heavens and travels swiftly around the Earth, while the Earth is far from the heavens and remains motionless in the center of the world." The Church simply could not support giving to Scripture "a sense contrary to the holy Fathers and all the Greek and Latin expositors" (ibid.: XII, 171–2; Galilei, 1957: 163).

Galileo certainly struggled to separate spiritual truth from scientific fact. Yet it should be remembered that he also attempted the even thornier task of trying to find Biblical evidence for new scientific discoveries. In a letter to Piero Dini on March 23, 1614, Galileo sought to employ the very text of Psalms 18 that Dini had described as "the greatest obstacle" to the Copernican system: "God hath set his tabernacle in the Sun" (ibid.: V, 301). In commenting on this passage and pointing out meanings "congruent" with the words of the prophet, Galileo used arguments that were typically Neoplatonic and Ficinian. An "animated, tenuous, and very swift" substance that is able to penetrate all places has its principal seat in the Sun. From this place it spreads throughout all the universe and giving heat and life and making fertile all living creatures. Light, created by God on the first day, is joined to the fecund spirit in the Sun, which for this reason is at the center of the universe, and from this place they begin anew to spread through the world. The Sun is "a conjunction in the middle of the world of the heat of the stars" and, as the source of life, Galileo compared it to the heart of animals that constantly renew the vital spirits (ibid.: V, 297–305).

Galileo was trying to demonstrate that there were indeed Biblical passages that revealed the Copernican depiction of the universe. For example, there was proof that the Sun lay at the center of the universe, and that its rotation on its own axis was the force that moves the planets. Galileo believed that the Psalmist knew this basic truth of modern astronomy: that the Sun "causes all other moving bodies in the universe to revolve around it" (ibid.: V, 319).

The moment that Galileo engaged his dialectical skills to find Biblical evidence of the new cosmology he compromised the value of his overall belief in strictly separating matters of faith and science, and distinguishing between the search for "how to go to heaven" and "how the heavens go" (ibid.: V, 319; Galileo, 1957: 186).

Hypotheses and Realism

The Council of Trent concluded in 1563, and the next year Galileo was born. Once the *Professio fidei tridentinae* was issued on November 13, 1564, the boundary between heresy and orthodoxy became unyielding. In 1592, Francesco Patrizi was condemned for having claimed the existence of a single heaven, the rotation of the Earth, astral life and intelligence, and the existence of infinite space beyond the sublunar sphere. Over a ten-year period, Pope Clement VIII banned Patrizi's *Nova philosophia*, Telesio's *De rerum natura*, and the complete works of Giordano Bruno and Tommaso Campanella, initiated proceeding against Giambattista Della Porta and Cesare Cremonini, and condemned Francesco Pucci to death, imprisoned Campanella, and had Bruno burned at the stake.

On December 20, 1614, from the pulpit of Florence's Santa Maria Novella, a Dominican priest named Tommaso Caccini declared heretical Copernican theory and the ideas of any who sought to correct the Bible. He railed against "the diabolical art of mathematics" and heretical mathematicians who should be banished from every Christian nation. In early 1615, after Galileo had already been formally denounced to the Holy Office for "suspicious and reckless" statements contained in his letter to Castelli, a publication appeared in Naples entitled *Letters of the Carmelite Paolo Antonio Foscarini on the opinions of Pythagoras and Copernicus*. That book argued that the Bible and Copernicanism *were* compatible. Cardinal Bellarmine's response to the piece was an extremely significant one. Foscarini and Galileo, wrote Bellarmine, would be wise to content themselves with speaking hypothetically. To say that "*assuming* the Earth moves and the Sun stands still saves all the appearances" better than the traditional system, not only "is to speak well" but "has no danger in it." However, to affirm that the Sun *actually* is fixed at the center of the universe and it is the Earth that moves "is a very dangerous thing, not only by irritating all the theologians and scholastic philosophers, but also by injuring our holy faith and making the sacred Scripture false" (ibid.: XII, 171; Galilei, 1957: 163).

Robert Bellarmine (1542–1621) was a Jesuit priest who was made cardinal by Clement VII in 1598. He was one of the most educated and influential figures in the Church at the time. His letter to Foscarini reiterated an idea that runs through the work of Simplicius, Giovanni Filopono, and Thomas Aquinas, namely that astronomy was pure "mathematics" and pure "calculation," and a construction of hypotheses for which it was not important to decide if they corresponded to physical reality or not. This same claim had been made by Osiander in his anonymous preface to Copernicus's *De Revolutionibus*, and was a notion that Bruno rejected with great vehemence. Even Kepler had said that Ptolemy's principles were "false" while those of Copernicus "true."

Galileo was in complete agreement with Bruno and Kepler on this score. He compared pure astronomy to philosophy, and argued for a physical description rather than a hypothetical description of nature. He believed that Copernicus's inquiry had not been a means for finding calculations that conformed to observation, but rather a discussion that involved "the composition of all parts of the universe *in rerum natura*" and "the true composition of all the parts of the world." He also felt that Copernicus never believed that the Ptolemaic world system corresponded to reality: "to believe that Copernicus did not truly believe that the Earth moves, in my opinion, is not to have read him [. . .]. Furthermore, it is my belief that he was incapable of moderation insofar as the central idea of his entire doctrine was based on the mobility of the Earth and the stability of the Sun: so, we either condemn it all or leave it in its entirety" (ibid.: V, 299).

Copernicus is Condemned

December 1615 found Galileo again in Rome, and once again in the mood for argument. He elaborated on the ideas in the Castelli letter in his *Letter to Madame Christina of Lorraine*. Also in the form of a letter, to Cardinal Alessandro Orsini, he wrote a *Discourse on the tides* which he later rewrote as the fourth day in the *Dialogue on the Two Chief World Systems*. However, all his plans and false hopes were soon interrupted. On 18 February, the theologians of the Holy Office examined Copernican doctrine in the crude version supplied to them by Caccini. The Holy Office declared the first proposition – "that the Sun is the center of the world and as a result, immovable of locomotion" – to be "foolish and absurd in philosophy and substantially heretical insofar as it expressly contradicts Holy Scripture." The second proposition – "the Earth is neither at the center of the world nor immovable, but moves" – seemed to merit "from a similar philosophical perspective, the same criticism as the first; and with regard to theological verity it is at least erroneous in the faith."

Pope Paul V ordered that Galileo be *warned* to abandon Copernican doctrine. Should he refuse, he would be *ordered* (or *commanded*), in the presence of witnesses and a notary, to renounce the censured doctrine and neither defend nor teach it. The distinction between a *warning* and an *order* would prove to be of critical importance because this was the basis for the charges against him of heresy in 1633, and his subsequent condemnation on those charges. On 26 February Cardinal Bellarmine summoned Galileo. The informal record of that meeting, which was not signed by the participants and appears to be in draft form, shows that Galileo was first *warned* by the Cardinal and then immediately afterwards (*successive et incontinenti*) *ordered*, in the name of the Pontiff and the entire Commissary of the Inquisition, to "totally abandon the propositions, and neither hold, defend, nor teach them in any way (*quovis modo*) orally or in writing." When Galileo was presented with this document at the second trial, he expressed surprise and said that the terms were "completely new [to him] and never heard of before." Many historians agree that the record of that meeting does not reflect what actually happened.

On 3 March, following Galileo's appearance before Bellarmine, the Inquisition issued a decree that banned the works of Copernicus pending correction. The decree also condemned and banned Father Foscarini's work, and any book in which Copernican doctrine was upheld. And so concluded the events put in motion by Lorini's charge against Galileo, who so far had not been personally implicated. Nor had his published works been banned. In May, after hearing about malicious insinuations and rumors that he had been forced to abjure, Galileo asked Bellarmine for a written statement. The cardinal attested that Galileo had never abjured or been punished: he had simply been notified of the Inquisition's decree that Copernican doctrine was contrary to Holy Scripture and therefore could be "neither defended nor held."

The Book of Nature

In 1623 Galileo published *The Assayer* (*Il Saggiatore*), one of the masterpieces of baroque literature; it was a biting and polemical work. The book was the offshoot of an argument with the Jesuit priest, Father Orazio Grassi, over the nature of comets. Grassi wrote the *Philosophical and Astronomical Balance* in 1619 as a response to the three lectures presented in the *Discourse on comets* by the "Galileist" Mario Guiducci. The Guiducci text was actually the work of the master himself. In both the *Discourse* and the *Assayer*, Galileo took the less-popular Aristotelian view of comets. Because parallax of the 1577 comet was much smaller than that of the moon, Tycho Brahe had correctly inferred that it lay beyond the moon. Galileo admitted that it was possible to measure distances by using parallax but rejected the application of the method to *apparent objects* (Galilei, 1890–1909: VI, 66). He classified comets as *optical phenomena*, like the sun's rays filtering through the clouds, and not *physical objects*.

In order to support his theory, Galileo attacked the astronomy of Tycho, who had believed the comets were physical bodies. Galileo, as we know, was hoping to cast the comets from the heavens and so demolish Tycho's reputation on Earth. He paid dearly for his attack on the greatest astronomer of his day: by casting himself in the role of the conservative Aristotelian, he was led to a host of inconsistencies (Shea, 1972: 88).

The Assayer, nonetheless, contains two of Galileo's most celebrated philosophical doctrines. The first stemmed from a series of reflections on the statement: "motion is the cause of heat." Galileo began by rejecting the notion that heat was an inherent quality of matter. The concept of matter or a bodily substance implied the concepts of form, relationship to other bodies, existence in time and place, motion or motionlessness, and contact or lack of it with another body. However heat, sound, odor and taste were not necessarily correlates of the concept of matter. If we were deprived of our senses, then human reason and imagination would never know that such properties existed. Though sounds, colors, odors, and tastes are considered to be objective qualities belonging to matter, they really are only "names," and when the live body is taken away, heat is reduced to a name. Galileo did not stop there. He stated that he was "inclined to believe" that what we perceive as heat is "a multitude of tiny corpuscles moving with great speed" and their "contact with our bodies is the sensation we call heat." Fire then has no other qualities than the form, multitude, motion, and penetration of fire-corpuscles and the sensation they produce.

So the physical world was laced with quantitative and measurable data, with space and "tiny particles" that move in space. Scientific knowledge can distinguish between what is objective and real in our world and what is instead subjective and produced by our senses. As Mersenne wrote in *Vérité des sciences*, the great modern abyss created between the physical universe and the world of sensory experience was far greater than the one imagined by the Skeptics.

Throughout the discussion of primary and secondary qualities, Galileo avoided using the term *atom*. He referred to *tiniest particles*, *smallest quanta*, *fire particles*, and *fire corpuscles*. At any rate, the reference was always to the smallest parts of a given substance (fire) and not to the ultimate components of matter. At the end of the *Assayer*, Galileo mentions "truly indivisible atoms." His references to atomistic–Democritean positions are especially important. In the first day of the *Discourse*, Galileo returned to the argument in the context of the phenomenon of cohesion. Simplicius condescendingly refers to "a certain ancient philosopher" and advises Salviati not to strike similar chords that "so differ from Your Lordship's well-tempered and organized mind, not only religious and pious but Catholic and holy."

The mention of particle theory in the *Assayer* did not escape the attention of the vigilant Father Grassi. In his 1626 response entitled *Ratio ponderum librae et simbellae* (*A Reckoning of Weights for the Balance*), he emphasized the similarities between the ideas of Galileo and those of Epicurus, who denied the existence of God and Divine Providence. To reduce sensory qualities to the level of subjectivity was to wage an open battle against the dogma of the Eucharist, a problem which Descartes too had to tackle. The problem lay in the fact that when the substances bread and wine are transubstantiated into the body and blood of Christ their external appearances – namely color, odor, and taste – are preserved. If, as Galileo suggested, these qualities were but "names, "then there is no need for the miraculous intervention of the Lord."

The second famous doctrine found in the *Assayer* expresses the author's belief that nature, though "deaf and unbending to our vain wishes" and producing its effects "in ways we cannot imagine," has an intrinsically harmonious order and structure, geometrical in manner: "Philosophy is written in this grand book – I mean the universe – which stands continually open to our gaze, but it cannot be understood unless one first learns to comprehend the language and interpret the characters with which it is written. It is written in the language of mathematics, and its characters are triangles, circles, and other geometrical figures, without which it is humanly impossible to understand a single word of it; without these, one is wandering about in a dark labyrinth (Galilei, 1890–1909: VI, 232).

The book of nature is written in characters quite unlike our own alphabet, and not everyone has the ability to read it. This idea was at the heart of Galileo's firm and unyielding belief that science should not be limited to formulating hypotheses – "saving appearances" – but could actually say something about the true constitution of the parts of the universe *in rerum natura*, and could represent the physical structure of the world. Immediately following the "book of nature" passage cited above, Galileo writes that he, like Seneca, seeks "the true constitution of the universe" and describes that search as "a large request and greatly desired by me."

Galileo's meaning was perfectly understood by those who feared and despised the idea that mathematical knowledge might uncover the objective structure of the world and so compare, in some way, with divine knowledge. Cardinal Maffeo Barberini (1568–1644) – elected Pope Urban VIII in 1623 – made his position on this question clear. Given that the explanation of a natural effect may be different from the one that seems best to us, all theories should be posited as hypotheses and remain there. Galileo states the exact opposite in the *Dialogue*, and argues that mathematical knowledge may be equal to divine knowledge. Simplicius the Aristotelian is struck by the "sharpness" of Salviati's statement: "*Extensively*, that is, with regard to the multitude of intelligibles, which are infinite, the human understanding is as nothing even if it understands a thousand propositions; for a thousand in relation to infinity is zero. But taking man's understanding *intensively*, in so far as this term denotes understanding some proposition perfectly, I say that the human intellect does understand some of them perfectly, and thus in these it has as much absolute certainty as Nature itself has. Of such are the mathematical sciences alone; that is, geometry and arithmetic, in which the Divine intellect indeed knows infinitely more propositions, since it knows all. But with regard to those few which the human intellect does understand, I believe that its knowledge equals the Divine in objective certainty" (Galileo, VII: 128–9; 1953: 103).

Scholars have often pointed out that Galileo's "philosophy" reflects a convergence of different traditions. Little is achieved by trying to determine whether Galileo was a Platonist or an Aristotelian; a follower of Archimedes or an engineer who managed to generalize specific experiences (Schmitt, 1969: 128–9). Galileo was indebted to each of those traditions; his view of the world as mathematically-structured was certainly linked to Platonism; the distinction he made between compositive and resolutive method was essentially Aristotelian; the mathematization of physics was derived from Archimedes; and the crafting and use of the telescope as well as his attitude toward the mechanical arts and the Venetian Arsenal was rooted in the intellectual tradition of the "superior artisan" of the Renaissance. What was more, he even used pseudo-Dionysian light metaphysics and Hermetic and Ficinian philosophy in his brief attempt to demonstrate that certain Copernican truths were revealed in Holy Scripture.

Galileo made use of each of these traditions. Mathematical idealism combined with the legacy of the "divine Archimedes," and a form of corpuscular theory, was destined to have an explosive effect on the western world.

The *Chief Systems*

The papacy of Urban VIII appeared to be a tolerant one. In 1626, three years after he was elected Pope, Urban VIII freed Tommaso Campanella and granted him a pension. It was in this new climate that Galileo wrote the *Dialogue on the tides* [*Dialogo sopra il flusso e il reflusso del mare*]. The title subsequently struck him as overly bold, and he prudently changed it to a seemingly more neutral one: *Dialogue concerning the Two Chief Systems of the World: Ptolemaic and Copernican* [*Dialogo sopra i due massimi sistemi del mondo, tolemaico e copernicano*]. The very title excluded consideration of the so-called "third world system" formulated by Tycho Brahe, which was generally favored in Jesuit circles.

Galileo demonstrated that he was observing the Pope's insistence on a hypothetical discussion in both the preface and conclusion to the *Dialogue*. In the preface he states, "I have taken the Copernican side in the discourse, proceeding as with a pure mathematical hypothesis," and then stated that the 1616 ban had been motivated not out of scientific ignorance but for reasons of piety and religion. For this reason the Earth had been declared motionless and the opposite theory dismissed as a "mathematical whim." Captious argumentation, a cautious preface, and the final reference to the Pontiff's "angelic doctrine" would not be sufficient to ultimately save Galileo from defeat and humiliation.

The bulk of the *Dialogue* differed substantially in tone from its cautious beginning. In the work, three men have a discussion in the Venetian palazzo of the aristocrat Giovan Francesco Sagredo (1571–1620). One of the interlocutors is Sagredo himself, who plays the role of the spirited and ironic freethinker. The second interlocutor is the Florentine Filippo Salviati (1583–1614), and he plays the part of the committed

Copernican, a scientist of strong convictions and rational mind. The third participant in the discussion is the fictitious Simplicius, the Aristotelian defender of tradition. Neither naive nor ignorant, he defends an order he believes is immutable and so fears all ideas which threaten it: "the aim of this kind of philosophizing is to subvert all of natural philosophy, and wreak havoc with heaven, earth, and all the universe." Salviati also represents the audience to which the *Dialogue* is addressed. As it was written in colloquial Italian rather than Latin, its readers were not meant to be "professors" like Simplicius, but members of the court, the upper classes, the clergy, and a new intellectual class. The discussion takes place over four days: the first is devoted to the destruction of Aristotelian cosmology; the second to the daily rotation of Earth; the third to the Earth's annual rotation around the Sun; and the fourth day to a *physical proof* of the Earth's motion by way of Galileo's tidal theory.

The *Dialogue* is not a book about astronomy in the sense that it does not explain a planetary system. It aims to prove the verity of the Copernican universe and illustrate why Aristotelian cosmology and physics are untenable. It never discusses planetary motion. Galileo gives a simplified explanation of the Copernican system without eccentrics and epicycles. Unlike Copernicus, he places the Sun at the center of circular orbits, and does not bother to account for observations of planetary movement. It has been correctly pointed out that Galileo had much more faith in his mechanical philosophy of uniform circular motion than in the accuracy of the measurements to which Kepler had long and patiently devoted himself. Again, this attitude was also responsible for his total neglect of Kepler's solutions to the problems of celestial kinematics (the theory of the ellipse dates to 1609 in *Astronomia Nova*).

The first day's conversation deals with the indefensibility of the Aristotelian "construction of the world." That world had a *double structure* based on the distinction between the incorruptible celestial world and the corruptible one of elements. Aristotle himself believed that "sensible experiments were to be preferred above any argument." And so Salviati points out to Simplicius, his reasoning would be more Aristotelian if he said that the heavens were changeable because his senses tell him so than to say they were not because Aristotle said it was so. While it was once impossible to directly observe the heavens, this obstacle has been overcome by the telescope. The peaks of the moon are not the only thing that force us to reject the traditional view of the universe. Although that view seems whole and stable, it is rent at its very center by faults and contradictions: it employs for example the perfection of circular motions to prove the perfection of the celestial bodies and then uses this latter notion to prove the perfection of those same motions. The qualities of generation and degeneration, divisibility and indivisibility, alterabilty and inalterabilty "are common to all worldly things, the celestial as well as the elemental." This statement is critical for it affirms that the heavens and the earth belong to the same cosmic system and share a single science of physics: a single science of motion explains the movements of both heaven and earth. The destruction of Aristotle's cosmology necessarily destroyed his physics as well.

The Destruction of Aristotelian Cosmology

The second day's conversation is entirely devoted to a detailed rebuttal of all classical and modern arguments against the motion of the Earth. Salviati suggests that the following phenomena would occur if the Earth moved: a stone dropped from the top of a tower would not fall at its base but at a point slightly to the west; a cannon ball shot to the west would range farther out than a cannon ball shot to the east; since we feel a strong breeze when we travel on horseback, so we should feel an east wind from the Earth's great speed at all times; and the buildings and trees on the surface of the Earth should be uprooted and cast off by the centrifugal force of terrestrial movement. As Galileo privately noted, "it is remarkable that we can urinate, when we rush so swiftly after the urine: or alternately, the urine should dribble along to our knees" (Galilei, 1890–1909: III, 1, 255).

Simplicius adopts an argument that had also been used by Tycho, and suggests that a stone dropped from the top of the mast of a ship at rest will fall in a perpendicular line, while a stone dropped from the top of the mast of a moving ship will fall in an oblique line and land towards the stern of the vessel at a distance from the base of the mast. The same phenomenon, presuming the Earth moves swiftly in space, should occur when a stone is dropped from the top of a tower. Simplicius, however, unwittingly misrepresents one fact – he has never made the experiment on the ship before – and at this point Galileo makes an important claim: anyone who tries the experiment will discover the opposite of what Simplicius says will happen. However, it is actually unnecessary to conduct the experiment because "even without experiment, I am sure the effect will happen that way." Using Salviati and Sagredo as his mouthpieces, he counters the arguments of the anti-Copernicans with the principle of the relativity of motion. Heavenly movement exists exclusively for observers on the earth, and so it is not absurd to attribute daily rotation to the Earth. Because the movement produces changes in appearances, the changes occur in the same way whether one assumes the Earth moves and the Sun does not, or vice versa. Whatever motion is ascribed to the Earth, by necessity "we who participate in it remain completely unaware of it." Salviati then suggests an experiment which alone refutes all the arguments against the motion of the Earth: imagine you are in a cabin below decks on a ship, and there are flies and butterflies in it with you, and a large bowl of water with fish in it, and a bottle that slowly drips water into another vessel beneath it, and if the ship moves at any speed "so long as the motion is uniform and not fluctuating this way and that, you will notice no change in any of the above objects, nor will you be able to tell if the ship is moving or standing still."

Galileo's assertion of relativity of motion had a significant consequence. In Aristotelian mechanics there was a crucial link between movement and the essence of objects. From this principle, not only could it be established which bodies were necessarily mobile and which were not, but it was also possible to explain why not all forms of movement were appropriate to all bodies. From the more open-ended Galilean perspective, rest and motion had nothing to do with the nature of objects; there was no such thing as an essentially mobile or immobile object, and it could not be determined *a priori* which bodies could move and which could not. The location of objects, in Aristotelian physics, was purposeful for both the object and the universe. Movement was represented as *motion* if it happened in space, as *alteration* if it related to qualities, and as *generatio* and *interitus* if it related to state of being. Motion was not a *state* but a process; a process in which things were created, realized, and completed. A body in motion changed not only in its relation to another body, but the body itself was subject to change. Galilean physics separated the motion of a body from the change that acts upon that same body. It signaled the end of both Aristotelian physics and medieval impetus theory which shared the notion that movement requires a *motive force* to produce motion and *conserve that motion* throughout. Rest and movement were now considered *continuous states* of a body. In the absence of external resistance, a force is required to stop a body in motion. Force produces not motion but acceleration. By overthrowing deeply-rooted concepts, Galileo laid open the way that would lead to the formulation of the principle of inertia.

Geometrization, Relativity, and Inertia

What textbooks describe as Galileo's principle of relativity (it is impossible to say whether a system is at rest or in uniform straight-line motion based on mechanical observations of the system itself) does not actually correspond to the theory he formulated for the purpose of demonstrating that it was impossible for an observer standing on the Earth to perceive the rotational movement of the Earth itself. Galileo enunciated a "broader" doctrine, according to which a motion "that does not fluctuate this way and that" which is common to all objects of a given system has no effect on the corresponding behavior of those objects and as a result can never be demonstrated within that system. In the famous example of the ship, Galileo used "non-fluctuating motion" to mean movement that was straight, direct, or moving along the same terrestrial meridian, and it is a distortion of the text to translate "non-fluctuating" as "rectilinear" (a term Galileo used several times elsewhere). The distinction is a significant one because the classical principle of relativity implies the concept of uniform rectilinear motion and acceptance of the principle of inertia (by which a body remains at rest or in uniform straight-line motion until an external force intervenes to change that state).

This concept is the very foundation of modern dynamics, and was never formulated by Galileo precisely because of the influence his cosmological ideas had on his physics. In the *Dialogue*, Galileo proposes a horizontal plane "sloping neither upward nor downward" upon which a moving object is indifferent to either acceleration or deceleration. Given a push, that object would continue to move the length of the plane and "if such a space were unbounded, the motion on it would likewise be boundless, that is, perpetual." The surface suggested by Galileo was not a horizontal plane tangential to the Earth's surface but a plane "which in all its parts is equally distant from the center of the Earth." Here Galileo means a spherical surface: "A surface that slopes neither up nor down would necessarily be in all its parts an equal distance from the center. However, is there any such surface in all the world? . . . none except our planet Earth, if it were given a good polishing."

Galileo's inspiration in this direction is revealed in the first day of conversation when he asserts the Aristotelian distinction between natural and unnatural motion. He claims that circular motion is *natural* and that constant rectilinear motion is *impossible*: "given that straight motion is by nature infinite, because the straight line is infinite and boundless, it is impossible for any moving object to naturally move in a straight line, that is, toward a place where it is impossible to arrive." We might imagine that straight motion perhaps occurred in the primordial chaos when the universe was still in disorder. Those straight motions, which upset the order of perfectly disposed bodies, might also "arrange in good order that which has been badly disposed." Straight motion may also serve "to move matter for the construction of a work; but once constructed, the object is to rest immovable, or if moveable, is to move only in a circle." After the most favorable and perfect placement of the parts of the universe, it is impossible for bodies to "still be naturally inclined to move in a straight line, for this would mean upsetting an object's own natural place, and causing disorder." We can then "imagine," as did Plato, that the planets in the beginning were given straight and accelerated motion, and only later, when they achieved a certain velocity, converted straight-line motion into circular motion "whose speed thereafter was naturally uniform."

Galileo is by no means simply making a literary concession to Platonic mythology. He reprises and expands the argument later on when Salviati debates the characteristics of circular motion: "this being the motion that makes the moving body continually leave and continually arrive at the end, it alone can be essentially uniform." Acceleration occurs when a moving body approaches the point toward which it has a tendency and retardation occurs because of its reluctance to leave and go away from that point. Instead, in circular motion the moving body "is continually going away from and approaching its natural terminus, so the repulsion and the inclination are always of equal strengths in it, and this equality results in a speed which is neither retarded nor accelerated; that is, a uniformity of motion." This concept of "perpetual continuation" – not a natural feature of the infinite line – derives from this uniformity and from the fact that circular motion can be "perpetually maintained." The conclusion clearly summarizes Galileo's belief that only circular motion can naturally suit the parts of the perfectly ordered universe; straight motion is assigned by nature "to its bodies (and their parts) whenever these are to be found outside their proper places, or arranged badly."

Infinite straight line motion is by nature impossible because nature "does not move where it is impossible to arrive." This seductive phrase expresses one of the greatest obstacles that Galileo the Copernican could not overcome. He continued to believe that circular motion was the superior motion; one which requires no explanation (in the new physics, circular motion would be explained with the help of non-inertial force). Galileo's greatest achievement and legacy – the joining of physics to astronomy – was based on the concept of the *inertial nature of circular motion*. The age-old idea of perfect celestial motion continued to hold sway over Galileo's physics.

Although it is difficult to read Galileo's *Dialogue* without "seeing" its *Newtonian potential*, it is important to not fall into the trap of attributing to a work ideas which later emerged from it. The concept of inertia in Newton's first law of motion was long in the making, and was the fruit of contributions by Descartes and Newton to Galileo's revolutionary idea. William Shea has suggested that in order to arrive at Newton's first law from Galileo's work, four steps were necessary: inertia had to be recognized as a basic law of nature, considered to imply rectilinearity, generalized from the motion of the Earth to all motion that occurs in empty space, and associated with mass as a function of the quantity of matter. Descartes managed the first three but only Newton conceived of the fourth (Shea, 1972: 9).

The Tides

In the close to twenty years that passed between writing the short treatise on the tides and the *Dialogue on Two Chief Systems*, Galileo came to believe that his theory of tidal motion and its cause was the final *physical proof* of Copernican cosmology. Galileo made the assumption that the ebb and flow of the sea was caused by motions of the Earth: daily west-to-east rotation on its axis and annual rotation around the Sun from west-to-east. The *combination of these two motions*, according to Galileo, made it so that every point on the globe moves "in a progressive and uneven way" and "changes speed sometimes by accelerating and others by decelerating." All the parts of the Earth therefore are subject to "motion that is notably different" even though no irregular or uneven motion has been assigned to the Earth."

Scholars have long emphasized that the "falseness" of Galileo's explanations (according to which tidal motion occurs only once every 24 hours) was not demonstrated by later scientific developments. Instead this explanation is hardly compatible with the progress that Galileo himself had made in physics and astronomy. After having introduced the classic principle of relativity to physics, Galileo (as Ernst Mach has noted) illicitly integrated two different frames of reference. The entire discussion on the second day of the *Dialogue* is aimed at proving that everything that happens on a moveable Earth happens just as it would on an immovable Earth. Why do only the seas reflect changes in the speed of the Earth? And why not all those objects not firmly attached to Earth? By the fourth day of *The Dialogue*, the planet Earth is no longer depicted as an inertial system (Clavelin, 1968: 480).

Galileo was trying to solve the question of the tides exclusively in terms of movement and the composition of movements; he rejected any notion of "lunar" influence and insisted unequivocally on mechanics. This insistence was paradoxically based on a strong aversion to the doctrine of influence and occult properties. Galileo was persuaded to reject as patently meaningless any tidal theory that referred to the "attraction" between the large bodies of water and the moon. He neither considered it as an alternative idea nor deemed it incoherent and false based on observable phenomena: he simply "discarded" it as the manifestation of an occult mentality. Galileo has the character of Sagredo dismiss the need for refuting similar trifles. That the Sun or the moon have anything to do with tidal motion is something "repugnant [. . .] to my intellect [. . .] which refuses to resort or succumb [. . .] to the attraction of occult qualities and similar useless fantasy." Galileo also expresses his great surprise at the fact that a man like Kepler, with such an "open and acute mind" and who believed in Copernicanism and had "at his fingertips the motions attributed to the earth" could have "given [his] assent to the moon's dominion over the waters, and to occult properties, and such puerilities."

Galileo's Tragedy

The controversy surrounding the *Assayer* had alienated Galileo from the Jesuit community. His enemies had no trouble convincing Urban VIII that Simplicius' reference in the *Dialogue* to "angelic doctrine" (according to which the causes of natural phenomena may not always be what is obvious, so it is wise to discuss them in terms of scientific hypothesis) was intentionally mocking of the Pope's authority. The Inquisitor of Florence ordered that sale of the *Dialogue* be suspended, and on October 1, 1632 Galileo was summoned to Rome by the Congregation of the Holy Office. Galileo managed to delay his departure until the first of the new year, but left on 20 January after being threatened to be taken there "bound in chains." He was quarantined because of a plague in Ponte a Centina, and after a lengthy delay finally arrived in Rome on February 13, 1633. On 12 April, a physically and morally debilitated Galileo appeared at the Holy Office. He learned that his crime was not that of having written the *Dialogue* but of having fraudulently obtained the license to publish it without informing the publisher of the 1616 injunction prohibiting the teaching and defense *quovis modo* of Copernican doctrine. Throughout the trial, Galileo maintained that he had been notified by Bellarmine and that the Cardinal himself – at Galileo's request – had later drawn up a document attesting to this. He stated that he did not recall an *injunction* issued in the presence of witnesses, and ended by stating that his true intention in writing the *Dialogue* was to prove that Copernicus' "reasoning" was invalid and inconclusive. This final phrase, made out of fear, delivered Galileo into the hands of the judges and eliminated any real possible defense. The consultors of the Inquisition had no difficulty showing that he was trying to trick the judges: "not only does he support Copernican theory with arguments never before heard, but he does so in the Italian language . . . so that his error has the best chance of catching on with the ignorant masses." Galileo moreover had attempted to step beyond the professional boundaries set for mathematicians: "The author claims to have presented a mathematical hypothesis but he gives it a physical reality, something a mathematician would never do."

In Galileo's written testimony, he states (10 May) that the terms of the 1616 document were "completely new" to him. A month later and following a second hearing the Inquisition delivered its verdict. On that same day, June 22, 1633, Galileo knelt before the cardinals of the Congregation and made a public abjuration: "I sincerely and in good faith denounce, curse, and abjure the above-stated error and heresies . . . and swear that in the future I will neither assert nor pronounce, orally or in writing, such things for which I could be again suspected, and if I know of other heretics or suspected heretics I will report them to this Holy Office" (Galilei, 1890–1909: XIX, 406–7).

The sentence was signed by seven of the ten judges and represented more than just a strike against Galileo and all his hopes and ideas. It also dealt a mortal blow to supporters of the new astronomy, within the Catholic Church itself, who believed that the Church could exercise a positive role in shaping culture. There is no doubt that the year 1633 was a critical one in the history of ideas and the history of science. Just a few months after Galileo's conviction (January 10, 1634), Descartes wrote to Mersenne – upon receiving the news – and urged him to postpone publication of his treatise on the world. He adopted the motto *bene vixit qui bene latuit*, or "he who hides well lives well," and confessed that he was tempted to burn all his papers. Ten years later, John Milton described his 1639 meeting with Galileo in *Areopagitica*: Italian intellectuals "were complaining of the enslaved state of science in their country; it had extinguished the lively Italian spirit, and for many years everything that was written was sycophantic and banal."

Galileo was sentenced to formal imprisonment. On July 1, 1633 he was transferred to Siena where he was welcomed as a friend by Archbishop Piccolomini. In December, he was allowed to return to his villa in Arcetri, outside of Florence, where he was ordered to live privately and neither "dine nor converse with" many people. His favorite daughter, Sister Maria Celeste, died on April 2, 1634 and Galileo wrote to a friend that he felt "immense sorrow and melancholy: loss of appetite, self-hate, and I continually hear calls from my beloved daughter" (ibid.: XVI, 85). At the end of 1637 Galileo began to experience loss of sight: "this earth, this universe, which I, by my remarkable discoveries and clear demonstrations had enlarged a hundred times beyond what has been believed by wise men of past ages, for me is from this time forth shrunk into so small a space as to be filled by my own sensations" (ibid.: XVII, 247).

The utterly ahistorical image which dominated nineteenth-century historiography and depicted Galileo as a freethinker and *ante litteram* positivist has been eclipsed. By the same token the many, at times clumsy, attempts to re-evaluate and justify the charges against Galileo have also been abandoned. On November 30, 1979, Pope John Paul II announced to a gathering of the Papal Academy of Sciences on the centennial of Einstein's birth, that Galileo Galilei "had suffered greatly [. . .] at the hands of the men and institutional organs of the Church" and declared that in his opinion a "wrongful act had occurred," of the sort that Vatican II already condemned (*Acta*, 1979: 1464).

The New Physics

Studies of Galileo since the 1970s have clarified not only the importance of his early works, *De Motu* and *Mechaniche*, but also demonstrated through a detailed analysis of passages in those works that the principal problems of Galilean physics date back to the decade between 1600 and 1610. In other words, Galileo's greatest scientific work had a very long gestation period. The *Discorsi intorno a due nuove scienze/Discourses on Two New Sciences* was published in Holland in 1638, technically without his knowledge. This work featured the same three characters that appeared in the earlier *Dialogue*. On the first two days, a true discussion on the strength of materials takes place. On the third day, the problem of naturally accelerated and uniformly accelerated motion is discussed, and day four deals with the paths of projectiles. Salviati reads aloud a Latin treatise on motion that has presumably been written by his friend, Academic. The reading of the treatise is only occasionally punctuated by questions from the two interlocutors. A "fifth day" (addressing Euclid's theory of proportions) and "sixth day" (on the force of percussion) were published later, in 1774 and 1718 respectively.

A new science was born with Galileo's theory of the strength of materials: an organic body of theory was for the first time applied to civil and military engineering and the science of building. In this context, Galileo's suggestion at the beginning of the *Discourses* is very significant; that is, that "philosophizing" should take into careful consideration the work of engineers and artisans. Sagredo declares that his conversation with "extremely skilled and eloquent" mechanics helped him several times in the course of his research into "hidden and almost unforseeable" effects. Galileo emphasizes first the importance of the *scale* of a structure as a determining factor of its strength, and shows why a model is relatively stronger than the full-scale structure. Prisms and cylinders that differ in length and thickness have a resistance to fractures that is directly proportional to the cube of the diameter of their base and inversely proportional to their length. The bones of a giant would have to be disproportionately thick for their length: the size of a structure cannot grow indefinitely in either art or nature. The cohesive strength of solids and the strength of materials can be explained by their particulate or atomic composition, and either the resistance to the formation of a vacuum between those particles (demonstrated by the resistance to separation of two smooth surfaces in contact) or else the presence of a viscous substance between them. However, in studying beams and fractures, Galileo overlooked the so-called effect of compression and mistakenly believed that the strength of beams was absolute.

Philosophers and historians of science have long studied the development of Galileo's theory of uniformly accelerated motion in the third day of the *Discourses*. The formulation of that theory was the culmination of a process that grew progressively more abstracted from experiential or qualitative elements. His earlier *De Motu* shows traces of ideas such as heaviness as a natural property of bodies, the natural descent of heavy bodies, and *vis impressa* as a temporary lightness that prevails over natural gravity. The speed of fall was considered in relation to the density and weight of the falling object. In the *Discourses* Galileo instead applies a purely kinematic analysis to the question of cause: velocity is directly proportional to the space that is traveled. This initial hypothesis was later discarded in favor of velocity as directly proportional to time, a considerably less obvious idea: "If an object at rest falls with uniformly accelerated motion, the distances traversed in any given time . . . are proportional to the squares of the times elapsed."

The relationship according to which D is proportional to T^2 (Theorem II, second assumption) is derived from Theorem I according to which the time it takes an object starting from rest to traverse a given distance while moving with uniformly accelerated motion is the same as the time it takes for the same distance to be traversed by the same object moving at a constant speed equal to half of the final and greatest degree of velocity attained by the previous uniformly accelerated motion. In the figure, AB represents the time elapsed during the uniformly accelerated motion of an object starting from rest through the space CD. EB represents the final and greatest degree of velocity attained during the time period AB. We then draw AE; lines that are equidistant and parallel to EB represent the increasing velocities after the initial moment A. Make F the midpoint of EB and draw FG parallel to AB and AG parallel to FB. The area of rectangle AGFB is equal to the area of triangle AEB because GF cuts AE at the intermediate point I. Extend the parallel lines contained in triangle AEB to GIF and "the sum of all the parallels in the triangle AEB is the same as the sum of all the parallels in the rectangle ABFG." The sum of the parallels in the triangle represents the "increasing velocities" of a uniformly accelerated motion while the sum of the parallels in the rectangle represents the velocities of an object moving at a constant speed. The sum of the velocities of each motion is equal: if velocity increases uniformly from zero to EB, the distance traversed is the same as that traversed in an equal time at the uniform speed IK (which is half of speed EB). To put it in non-Galilean terms, the sum of the increasing instantaneous velocities in the accelerated motion is equal to the sum of the uniform instantaneous velocities at the mean speed IK.

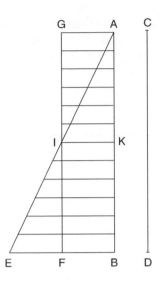

Galileo did not fully identify distance with area, and his knowledge of infinitesimal calculus was insufficient to allow him to state "that the sum of an infinity of little lines, each representing a velocity, adds up to something different, a distance" (Shea, 1972). Infinitesimal calculus would prove to be the mathematical method required for dealing with continually varying quantities.

The problem which Galileo addressed in the brief Latin treatise inserted in the *Discourses* was that of finding a definition of uniformly accelerated motion that "exactly described [. . .] the kind of acceleration that nature gives to falling objects." Galileo declared that he was practically "led by the hand" to his definition because nature is served, in all its works, "by the most immediate, simple and efficient" means. A stone that falls from rest progressively gathers speed, so why not imagine that the increase in velocity occurs in the most simple and obvious way (*simplicissima et magis obvia ratione*)? There are two possibilities that meet the requirements of an increase or increment that "always occurs in the same way": velocity might be proportional to distance or to time. It has often been pointed out that the choice made by Galileo (who perceived the two options as equally simple) was tied to his erroneous demonstration of the logically contradictory nature of the alternative hypothesis.

"Through a similar uniform subdivision of time we understand that increments in velocity occur with the same simplicity." This is possible because we establish in the abstract (*mente concipientes*) that "a motion is uniformly and continuously accelerated when it receives equal increments of velocity in equal times." Sagredo observes that the definition is arbitrary – "conceived of and allowed in the abstract" – and it is doubtful whether it can be adapted to reality and physically proven. Simplicius makes the same objection at the conclusion of the lengthy demonstration. While persuaded by the validity of the demonstration, he doubts that nature really applies that sort of motion to falling bodies: "it seems to me, and to others like me, that it would be opportune in this case to gather some experiences." And so Galileo responds to this request with his famous description of a perfectly spherical, hard and smooth bronze ball that rolls down a smooth, hard, straight inclined plane. The formulation of the principle was not based on this experiment though, and Galileo states it clearly on that same page: the experiment was carried out "to demonstrate that naturally falling bodies accelerated according to the above proportions."

The Fourth Day of the *Discourses,* devoted to projectile motion, is an example of the exceptional quality of Galilean science. In these pages Galileo shows that the trajectory of a projectile is parabolic and a combination of two independent motions that *do not interfere with one another*: a forward uniform motion on the horizontal and a downward uniformly accelerated motion on the vertical. This law, which combines the principle of inertia and the law of free fall, allowed Galileo to determine speed, height, range, and quantity of motion. Galileo effectively changed the way people understood motion and introduced a radically different way of relating motion and geometry.

Galileo continued to write letters, and study and discuss problems late into his life. With the help of Viviani and his youngest disciple, Evangelista Torricelli, he was even able to recapture some of his former enthusiasm: he argued with Fortunio Liceti, followed the debate between Viviani and Torricelli, and clarified his views on Aristotelianism. In the pre-dawn hours of January 8, 1642, the now blind eyes that had been the very first to see new stars and the peaks and valleys of the moon, closed once and for all. To avoid "scandalizing good people" it was deemed unfit to construct "an august and sumptuous resting place" for his mortal remains. The Pope's nephew wrote that it would be inappropriate "to build a mausoleum for the body of one who had been convicted by the Tribunal of the Holy Inquisition and died serving his sentence."

7

Descartes

A System

One of the reasons for the extraordinary success of Descartes' grand construction was that it was introduced to European culture as a *system*. A system founded on reason, with no mention of the occult and vitalism, and seemingly capable of linking (differently from medieval Scholasticism) the science of nature, natural philosophy, and religion. Last but not least, it offered men a coherent, harmonious, and complete picture of the world in a doubt-ridden era of intellectual revolution.

The acceptance and spread of the Cartesian system was slow, painstaking, and controversial. Already banned by the universities of Utrecht and Leiden in the 1640s, Cartesian philosophy was condemned in 1656 in all the Netherlands by decree of the Synod of Dordrecht. In 1663 the Catholic Church listed Descartes' published works on the Index of prohibited books. In Italy, the philosophy of Descartes went together with that of Gassendi and Bacon, and the legacies of Telesio, Campanella, and Galileo. Tommaso Cornelio "introduced the work of René Descartes to Naples." In his 1681 *Ideas on the uncertainty of medicine* (*Parere sull'incertezza della medicina*) Leonardo di Capua proposed that Cartesian science be added to Galilean science. Michelangelo Fardella of Trapani taught Cartesian philosophy at the University of Padua from 1693 to 1709.

By the late seventeenth century, all of the great universities of Europe had accepted Descartes' method and the bans against it had lapsed. For the entire second half of the seventeenth century, Cartesian philosophy and physics dominated European intellectual life. Hobbes, Spinoza, Leibniz, and later, the great Enlightenment philosophers, measured their work against that of Descartes; its greatest critics, from Locke to Vico, also compared their work to his. The intense debate between Cartesianism and Newtonianism only ended – with the defeat of Cartesianism – around 1750.

"I Come Forward Masked"

René Descartes (in Latin Cartesius) was born on March 31, 1596 in the town of La Haye (now known as La Haye–Descartes) in the Touraine. His family belonged to minor nobility. His mother died when he was a year old, and he was raised by a wet nurse and his maternal grandmother. At about the age of 9 or 10, he was sent to the prestigious Jesuit college at La Flèche where he remained for eight years. Despite years of earnest study and avid reading, upon leaving school in 1618 he nevertheless felt "disconcerted by doubts and errors" and believed his school years had only helped him "discover [his] own great ignorance." At this time Descartes resolved to "seek no science other than the knowledge of myself, or of the great book of the world," and spent the remainder of that year "traveling, visiting courts and armies, meeting a great variety of people, and gathering various experiences." He joined the army of Prince Maurice of Nassau in Breda, Holland. It was here at the end of 1618 that he met Isaac Beeckman (1588–1637). Beeckman taught at the Latin school in Dordrecht, and was a vastly well-read man of encyclopedic knowledge. His famous *Journael* contains notes throughout of the ideas and thoughts (many of them original and important) prompted by his vast reading and personal research. Descartes gave the manuscript of the *Compendium musicae* (eventually published after his death) to Beeckman as a gift. In this early work Descartes had already begun to study nature in mathematical terms. In 1619 Descartes joined the army of the Bavarian Elector. On the night of 10 November in the vicinity of Ulm, Descartes had a mystical-scientific experience during which "the foundation of a wonderful science" was revealed to him. The next day he pledged to make a pilgrimage to the shrine of the Madonna of Loreto as soon as his project was completed. There has been speculation that he joined (or was associated in some way with) the Order of the Rosicrucians. While we have no evidence for this, Descartes was undoubtedly attracted to the eschatological and millenarian aspects of the order of the followers of the mysterious Rosenkreutz, who, according to the 1615 Rosicrucian publication, *Confessio*, was born in 1378 and lived for 106 years.

Following trips to Bohemia and Hungary, Descartes returned to France in 1622 and the next year went to Italy. Between 1627 and 1628 he most likely drafted *Regulae ad directionem ingenii* (*Rules for the Direction of the Mind*) the important predecessor to his method. In 1629 Descartes moved to Holland where he lived until 1649. In 1630 he began *Le Monde ou Traité de la lumière* (*The World*) but decided against publishing it when he learned in 1633 of Galileo's condemnation by the Inquisition. It first appeared in print in 1664, more than 14 years after his death. *A Discourse on Method* (*Discours de la méthode*), one of the cornerstones of modern philosophy, was published in Leiden on June 8, 1637 as the introduction to three scientific essays: *Dioptrique*, *Météores*, and *Géométrie*. The *Dioptrique* contained the law of refraction, also known as the law of sine, according to which as a ray of light passes from one medium to another there is a fixed ratio between the sine of the angle of incidence and the sine of the angle of refraction. The collection, however, the fruit of over two decades of labor and Descartes' carefully chosen representation of himself as an intellectual, had a strange fate. It was split up so that until 1644 the *Géométrie* – the most discussed and debated work of the seventeenth and eighteenth centuries – was read separately, and the same thing happened later to the *Discourse*, which was being read exclusively as a "philosophical" piece. *Meditationes de prima philosphia* (the *Meditations*), a metaphysical treatise begun in 1629, was published in Latin in Paris in 1641 together with a series of "objections" and "replies." In 1647 it was translated into French. In 1642 Cartesian doctrine was condemned by the University of Utrecht, and in 1643 Descartes responded with *Epistola ad Gilbertum Voetium* (Gijsbert Voetius was one of his principal critics and accusers). The *Principia philosophiae* (*Principles*) was published in 1644; the last three books in the collection explain his physics. In 1647 the University of Leiden accused Descartes of Pelagianism. He went twice to France, and then in 1649 moved to Stockholm at the invitation of Queen Elizabeth of Sweden. He published *Traité des passions de l'âme* (*Passions*) that same year. Descartes died of pneumonia in Stockholm in 1650.

Descartes was in large part personally responsible for his reputation as a solitary philosopher who read few books and listened instead to the inner voices of his consciousness. However, his vast correspondence (many letters of which discuss important scientific ideas) alone is enough to lay the myth to rest. Descartes had read all of the work of the leading intellectuals of his time: Simon Stevin and François Viète on algebra and mathematics; Kepler and Christoph Steiner (1575–1650) on optics; Gabriel Harvey on medicine; Francis Bacon on natural philosophy and method. He was familiar with ancient Greek mathematics and the sophisticated versions of it in the manuals of Christoph Clavius (1537–1612). He knew about Arabo-Latin optics, and the physics of the modern-day atomists. He remained, on the whole, faithful to his youthful musing: "I am now about to mount the stage, and I come forward masked." Descartes has been described as a reluctant revolutionary, eager to avoid confrontation with the official philosophy and successful in doing so without ever compromising his own point-of-view (Shea, 1991: 112).

Mathematical Equations meet Geometry

Modern science was not founded on the ability to generalize empirical observation but, as we have seen (in the case of Galileo for example), on the analytical ability to abstract; to abandon the realm of common sense and sensory experience. What fundamentally revolutionized physics was, as we have discussed, its *mathematization*, and Galileo, Pascal, Huygens, Newton, and Leibniz all made important contributions. However, it is Descartes whom we find at the heart of this grand and complex process.

Analytical geometry departed significantly from Greek geometry with Descartes, who based his work on the late sixteenth-century discoveries of François Viète. The old geometry solved any arithmetic and algebraic problem in terms of geometry. Descartes demonstrated the possibility of reducing geometrical problems to algebraic ones. From the beginning of his 1637 treatise *Géométrie*, he discussed "introducing arithmetical terms to geometry" and radically departed from the tradition that associated squared or cubed algebraic quantities to "like" geometric quantities, with the "power" of a number equal to a "number of dimensions." In other words, for Descartes $(a + b)^2$ – the square of the sum of two lines – was itself a line and not an area. Quadratic or cubic equations correspond to linear geometric entities. He assigned letters to the lines of a geometric figure, and the solution to an equation formed by them is the length of an unknown line. Moreover, the introduction of coordinates, today still referred to as Cartesian, made it possible to both define the position of a point and assign (kinematically) an equation defining a straight line or a curve drawn from that point. The equations can be represented in geometric terms and the curves by means of equations. The properties of a given curve can be studied by applying algebraic operations to the equation that represents it.

Physics and Cosmology

Descartes' "discovery" meant that the problems of physics, and in particular of mechanics, could now be subjected to the precision of algebra. Let us consider, for example, the definition by equation of the trajectory of a projectile. Ernst Cassirer described it well when he wrote that space, time and velocity, which when taken separately do not seem easily related to one another, become instead homogeneous; mathematics had discovered a procedure by which the unit of measurement of one quantity can be referred to that of the other.

In his grand attempt to completely and rationally *reconstruct* the physical world, Descartes arrived at an important definition of the concept of movement and clearly formulated the principle of inertia. His second "law of nature" states that "all motion is in itself rectilinear" (Descartes, 1967; 1985: 96). Contradicting both Copernicus and Galileo, Descartes states that "each of the individual parts of a body in motion does not move along a curved path but along a straight line" and that "any body that moves is compelled to move along a straight line and not a circular one." In circular motion there is a tendency to "recede without respite" from the circle described: "you can even feel it in your arm as you swing a stone in a sling." This "consideration" struck Descartes as significant, and helped to finally lay to rest the myth of circular perfection. At the same time, in 1629 Descartes based the law of falling bodies on the incorrect formula that the velocity of the body in motion was a the function of the distance traveled and not the time elapsed.

Descartes' idea of motion differed substantially from the traditional one. Motion was not a *process* but a *state*, the ontological equivalent of rest. To be at rest or in motion caused no change whatsoever in a body. There was nothing in the universe but motion and matter, and Cartesian physics was strictly mechanistic: all forms of inanimate objects could be described by attributing nothing more than movement, size, form, and the composition of parts. *Res cogitans* and *res extensa* were two strictly separate realities. Nature was neither psychic nor could it be interpreted in animistic terms. In the *World*, Descartes described nature in these terms: "By 'nature' I do not mean some goddess or any other sort of imaginary power, rather I use this term to signify matter itself taken together with all the qualities I have attributed to it, and under the condition that God continues to preserve it in the state in which he created it." From the mere fact that God continues to preserve it, the many changes in its parts cannot be attributed to the action of God but to nature itself: "The rules by which these changes take place I call the laws of nature."

In the tradition of mechanics, Descartes used models to explain nature: because an idea is not a reflection of the real world, there is no reason to believe (even though we are usually convinced of it) "that the ideas we have in our mind are in any way similar to the objects they depict." Just as words, whose meanings are determined by human convention make us think of things to which they bear no resemblance," so nature established "signs" which would give us sensations without having to be similar to the sensation.

Descartes defined *matter* as *extension* and considered all matter in those terms. The only difference between matter and the space it occupies is motion. A material object is a form of space that can be transported from one place to another without losing its defining properties: "extension in length, breadth, and depth defines both space and physical bodies; and the only difference between them consists entirely in the fact that we attribute a particular extension to a body that we imagine changing places with it every time it moves from one spot to another" (Descartes, 1967: II, 77). If *space* and *motion* make up the world then the Cartesian universe is the *realization of geometry*.

The Cartesian identification of space and matter implies a number of consequences: (1) the identity of universal matter; (2) the indefinite extension of the world; (3) the infinite divisibility of matter; and (4) the physical impossibility of the void. Like Euclidean space, the world or "the extended matter that makes up the universe is limitless" (ibid.: II, 84). Because God alone can be *infinite* and the finite human mind cannot understand *infinity*, "let us call these things *indefinite* rather than *infinite* so as to reserve that word for God alone" (ibid.: I, 39–40). Descartes was even more radical in his rejection of voids than Aristotle had been. He argued that void space was impossible for if it indeed existed, it would be an existing nothing; a contradictory reality. A void or vacuum has no properties or dimensions. The distance between two bodies has a dimension and the dimension is a matter too "subtle" to be perceived and which we only imagine to be a "void." Descartes believed that physical reality was corpuscular but divorced himself from ancient atomism for two reasons: he believed in the infinite divisibility of matter and rejected the existence of the void or vacuum.

In *Meteorology*, Descartes wrote that water, earth, air, and all the rest that surrounds us, is made up of "many particles that vary in form and size, particles that are never so well-arranged or connected as to avoid being surrounded by many little gaps; these are not voids but full of a fine matter through which light is transmitted" (Descartes, 1966–83: II, 361–2). Descartes inquired not only into the composition of the universe but also its formation. God had taken extended matter and divided it up into cubes, setting all the parts of the universe in motion so that the cubes were "violently agitated." The three compositive elements of the universe were formed in this manner. The resulting friction wore down the angles and edges of the cubes and they changed shape and became tiny spheres. The infinitesimal particles produced by the "abrasion" became the "fiery" *first element*, which is the matter of light. The first element is "like the finest and most penetrating liquid in the world"; its parts have no determined shape or size but "change shape every moment to adapt to fit the space they enter." No space is too small for these tiny particles to permeate and fill, and Descartes compared their motion to a rushing river flowing directly from the sun, causing the sensation of light (Descartes, 1897–1913: II, 364–5). If the first, or fire element, *is* light, then the *second element*, or *aether*, forms the heavens and transmits light. Its particles are all "more or less spherical and joined together like grains of sand or fine dust." Because this matter cannot be crushed or squeezed, the "first element easily slips in" to fill the little gaps between the aether particles. Like the others, the *third element* is also a product of the abraded matter that came together into threaded, screw-like particles and formed the Earth and the planets. The parts of the third element are "so large and firmly joined that they have the power to constantly resist the movement of other bodies." Water particles, on the other hand, are "long, smooth, and shiny like little eels, and despite the fact that they are joined and woven together, do not get knotted up or stuck together in such a way that they cannot be unstuck easily one from the other" (Descartes, 1966–83: II, 362–3).

The *fine matter* of which the heavens are composed plays a fundamental role in Cartesian physics: it is the basis for rarefaction and condensation, transparency and opacity, elasticity, and gravity. Since motion necessarily requires displacement or rearrangement in a completely full universe, then every movement tends to create a swirl or vortex. All movement in the universe is in some way circular: "which is to say that when a body leaves its place it always changes places with another one, which changes places with a third body, and so forth until the last body simultaneously occupies the space left empty by the first body so that no void exists between them when they move, just as there is no vacuum when they are at rest." Because the vacuum does not exist in the universe, it "was not possible for all the parts of matter to move in a straight line, but being roughly equal and with the same capacity to be deviated, they had to assume in unison a certain circular motion." Since God put them into different motion from the very start, they began to turn "not around a single center, but around many different centers." The globular particles of the second element formed large whirling vortices. As a result of centrifugal force the particles of the first element were pushed toward the center. The sun and fixed stars are globe-shaped masses of first element particles. Both the first and second element surround, in the guise of liquid vortices, the sun and stars. The planets, which are dragged around the sun by the motion of a smaller vortex, "float" in the liquid vortices like the flotsam in a stream that spins in little whirlpools while at the same time being dragged along by the greater current. Comets are not optical phenomena but actual celestial bodies that endlessly travel to the borders of the vortices, passing from one to another. In an indefinitely vast universe, the vortices are prevented from expanding by their proximity to one another. Finally, the vortices generate the force that holds the planets in their orbits. Though Descartes' doctrine ignored the technical aspects of planetary astronomy (Descartes never mentions Kepler's laws), it nevertheless respected the basic laws of mechanics: without resorting to "occult forces" of any type, it could explain the rotation of the planets around the sun.

In a completely full universe with no vacuums, all movement is necessarily the result of impact, and for this reason the idea of *impact* or *percussion* was central to Cartesian physics. Given the immutability of God, the quantity of motion in the universe remains constant. "Quantity of motion" was a term Descartes used to describe the product of a body's "size" and its velocity. However, Descartes intended "size" differently from the modern meaning of "mass," and did not treat velocity as a vector quantity (Westfall, 1971: 121). However, it is not necessary that the quantity of motion of every body remain constant. Motion can be transferred from one body to another at the moment of impact. The third law of nature emerged from this premise: "if a body in motion meets another stronger than itself, it loses none of its motion, and if it meets a weaker one which it can move, it loses as much of its motion as it gives to the other" (Descartes: II; Westfall, p. 122). According to this law, a body in motion cannot move another body it collides with if that body is at rest and larger than it. Galileo had clearly understood that no matter the size of a body at rest, any sized body that strikes it will always confer some motion. Only a body at absolute rest, that is of infinite mass, could avoid change as a result of impact. In *De motu corporum ex percussione* (written in 1677 but only published in 1703), Christiaan Huygens rejected Descartes' collision theory. In his personal copy of the *Principles*, Newton wrote the word "error" so many times that – according to Voltaire in the fifteenth of the *Lettres philosophiques* – "so tired of marking errors he just threw the book away."

The World as Geometry

Descartes once wrote to Mersenne that "all my physics is nothing but geometry." Like geometry, with which it is closely linked, Cartesian physics was based on a series of axioms and was rigorously deductive in nature. Koyré has pointed out that Descartes, unlike Galileo and Newton, did not ask how nature actually behaves but instead wondered how it *should* behave. The idea of physics as geometry, and the world as "geometrized," led Descartes toward an "imaginary" physics that was described as "philosophical fiction" by not only Huygens, a Cartesian, but also Newton and numerous other critics. In many cases, the connection to experience and the search for empirical proof of theories were but illusory in Descartes' method. The Cartesian laws of nature, according to Koyré, were laws *for* nature that nature could not help but conform to because those laws constitute nature itself.

No greater testament to the seductiveness of Descartes' construction exists than the eloquent words of Huygens (1629–95) in a letter to Bayle on February 26, 1693. He wrote that Descartes had found the way to make his guesses and inventions accepted as truths, and that readers of the *Principia philosophiae* experienced the same thing that readers of a good novel do every time they trade fantasy for reality. "When I read this book for the first time, I was carried away by it and certain that any difficulties I had were the result of my not understanding the author's thinking. I was then 15 or 16 years old [. . .] Today I can find almost nothing I believe to be true in either his physics, metaphysics, or meteorology" (Descartes, 1897–1913: X, 403).

Naturally, the autobiographical reflections of mature philosophers and scientists tended to simplify what were complex and nuanced intellectual events. Huygens had studied at The Hague and Leiden with Cartesian professors. In Paris and London he was introduced to the circles of Mersenne and the Royal Society. His work reflected the union of sophisticated mathematical and mechanical theory and an interest in machines and technology that link him to the Baconian and Galilean tradition. With the exception of the optical theory in *Traité de la lumière* (1690), Huygens essentially remained loyal to the Cartesian concept of mechanics. His anti-Newtonian position in *Discours sur la cause de la pesanteur* (1690) is the result of this background.

Unlike Huygens, Descartes used neither formulas nor mathematical terminology in writing his physics. Though his laws were not expressed in mathematical terms, his physics was mathematical. Cartesian "mathematicism" was manifested purely in the axiomatic and deductive construction of the world. The very title of Newton's *Philosophiae naturalis principia mathematica* (published in London in 1687) expressed his controversial opposition to Descartes and Cartesian physics. Newton described the principles of natural philosophy in mathematical terms and, at the same time, adapted the great lessons of Bacon's, Hooke's, and Boyle's experimentalism to serve his own purposes.

8

Countless Other Worlds

An Infinite Void

The writings of Giordano Bruno (1548–1600), an ardent Copernican charged with heresy by the Inquisition and burned at the stake in a public Roman square, were in great demand and avidly read all over Europe. Bruno's name became a symbol. He believed that the Copernican theory of the universe was not a purely mathematical one, like the "ignorant and presumptuous ass" Osiander claimed it to be in the preface he penned to *De Revolutionibus*. According to Bruno, Copernicanism was not simply a new astronomical system, but a new world view. It represented the triumph of new truths and, at the same time, was an instrument of liberation: "This is the kind of philosophy that opens the senses, delights the spirit, expands the mind, and brings man true mortal bliss."

The world of Copernicus was a closed and finite one. In the *Ash Wednesday Supper* (1584), Bruno not only refutes the classic arguments against the motion of the Earth but also firmly asserts the infinity of the universe: the universe is infinite, and there is no body that must be at its center or periphery or even in between." The infinite world – produced by an infinite Cause – is found in the infinity of space. He writes, "Let us call space infinite, for there is no reason, need, possibility, sense or nature that must limit it [. . .]. The Earth therefore is not the absolute center of the universe but the center of our region [. . .]. This is how God's excellence is magnified, the greatness of his kingdom made manifest; he is glorified not in one sun but in countless suns; not in one Earth or one world, but in 200,000, I say, in an infinity of worlds" (Bruno, 1907: 275, 309).

Bruno believed that motion and change were positive realities and rest and stasis synonymous with death. Only that which changes is alive, and perfection is found in creation and change. In *De l'infinito, universo e mondi* (1584), he writes: "There are no ends, boundaries, limits, or walls which can defraud or deprive us of the infinite multitude of things. [. . .] For from infinity is born an ever fresh abundance of matter" (ibid.: 274). On the same page Bruno discusses Democritus and Epicurus. The Copernican universe and other similar worlds belong to an infinite and homogeneous space that "we can freely call empty." The infinite void proposed by atomists such as Democritus and Lucretius became a sort of "natural home" for not only the Copernican solar system, but all the other numberless systems as well (Kuhn, 1957: 237). The animate universe that Bruno described has even been termed *astrobiology*. Bruno did not limit himself to describing celestial spheres and epicycles as "a poultice or prescription for healing nature [. . .] in the hands of Master Aristotle"; but went on to reject circular uniform celestial motion and the idea of movement that is "continuous and regular around a center"; he claimed that in the physical universe there was no such thing as perfect motion and perfect form. Bruno believed that the laws governing celestial motion had to do with the individual stars and planets that had "souls of their own" that guided them through the heavens: "Motion is intrinsic to the very nature, soul, and intelligence of these racing objects."

Bruno's cosmology clearly distinguished between *universe* and *world*. A world system was not the same thing as a system for the universe. While the science of astronomy was a legitimate and plausible science for the world man could perceive with his senses, it could not be applied to what lay beyond this realm: an infinite universe filled with those "great animals" that we call the stars, and which contains an infinite number of other worlds. This universe has no size, dimensions, form, or shape, nor is it ordered or harmonious. There can be no *system* for a universe that is at once both uniform and formless.

Tommaso Campanella (1568–1639) wrote the impassioned *Apologia pro Galilaeo* in 1616 while serving a prison sentence begun in 1599. He argued that there was a profound difference between suggesting that many different worlds make up a *single system* and suggesting that a plurality of lost and disordered worlds exist in infinite space. Galileo, with the help of his impressive instruments, was able to show us stars hitherto unknown and help us learn that the planets were like the moon, receiving their light from the sun and rotating one around the other. From Galileo we learned that the transmutation of elements takes place in the heavens, that clouds and vapors exist among the stars, and that there are a great number of worlds. Campanella's reply to the ninth *Argumenta contra Galilaeum* explains that "from Galileo's opinion it follows that there are many worlds and earths and seas." Brother Tommaso also explained that Galileo's theories were not to be confused with those of Democritus and Epicurus, for Galileo had said that all the celestial systems constitute one world enclosed in an immense heaven and coordinated into a greater totality: To suggest that there exist many worlds without any order as a whole, as Democritus and Epicurus had done, is an error in faith because it follows that such worlds are formed randomly and without the intervention of God. On the other hand, to suggest that there is a plurality of lesser worlds within a greater and divinely ordered system does not contradict Scripture but only Aristotle (Campanella, 1994: 111).

The whole of Bruno's thinking was based on the idea of many worlds not coordinated into a single one. Copernicus, Kepler, Tycho, and Galileo each had solidly upheld the picture of a single, ordered universe which, in their eyes, was the expression of a divine order; the manifestation of a mathematical-geometrical archetype or principle. Galileo wrote that "I wholeheartedly agree with and confirm what [Aristotle] has said to this point, that the world is a body endowed with every dimension, though an absolutely perfect one; in addition, I suggest that as such, it is by necessity in perfect order, that is, its parts make up a whole and are perfectly arranged with respect to one another" (Galilei, 1890–1909: VII, 55–6). This is a radical alternative to Bruno's view of the universe. Bruno's unique mixture of Cusa's Platonism and Lucretian materialism produced an image of a 'haphazard' universe; an image rejected not only on grounds of impiety but also because it challenged tradition and was not accepted by the theorists of the new astronomy.

An Infinitely Populated Infinite Universe

In *The Great Chain of Being* (1936), Arthur O. Lovejoy – the original historian of ideas – laid out the five "revolutionary theses" upon which the late seventeenth- and eighteenth-century cosmography was based. These were: (1) the assumption that other planets in our solar system were inhabited by living, feeling, thinking creatures; (2) the shattering of the external walls of the medieval universe, whether these were the outermost crystalline sphere or the definite region of the fixed stars, and the dispersal of these stars throughout vast and irregular space; (3) the belief that the fixed stars were suns similar to our sun, all or almost all surrounded by planetary systems of their own; (4) the supposition that the planets in these other systems had conscious inhabitants; and (5) the assertion of the infinity of the physical universe in space and of the number of solar systems in it (Lovejoy, 1936: 108).

Not one of the above ideas can be found in the writings of Copernicus. Both the doctrines of the infinite universe and the multiplicity of worlds were rejected in different ways by the three greatest astronomers of Bruno's age: Tycho Brahe, Johannes Kepler, and Galileo Galilei.

Kepler unequivocally rejected Bruno's vision of an infinite universe and rejected the comparison of the sun to the fixed stars. He upheld the unique and "exceptional" character of the solar system, contrasting it with the immovable mass of the fixed stars. As to whether the fixed stars were all located on a single spherical surface and therefore the same distance from Earth, Kepler was unsure of the answer. He nevertheless believed that at the "center of the universe lies an immense void or great cavity surrounded by a row of fixed stars that define and enclose the universe like a wall, and our Earth and sun and the moving stars are located inside the great void" (Kepler, 1858–71: VI, 137).

Kepler rejected Bruno's infinite universe both *before* and *after* Galileo's telescopic discoveries. The universe was created by God the geometer who gave it a geometric design: the vast void is the same as *nothingness* and the fixed stars were not haphazardly scattered throughout space. He wondered "how it was possible to find the center in an infinite space, which in infinity might be anywhere? Every point in infinity is equally, that is infinitely, separated from the infinitely distant endpoints. The resulting paradox – that a single point is both a center and not a center – and many other contradictory things are most correctly avoided by one who, just as he finds the sky of the fixed stars limited from the inside, limits it also on the outside" (Kepler, 1858–71: II, 691; cf. Koyré, 1957: 70). The solar system remained unique in the universe. There were two possible explanations for Galileo's discoveries: the new (fixed) stars observed by Galileo are not seen by the naked eye either because they are too far or else because they are too small. Kepler firmly chose the second explanation (Koyré, 1957: 76).

Kepler's goal in the *Dissertatio cum Nuncio Sidereo* (1610) was to show that Galileo's astronomical discoveries did not prove the validity of Bruno's infinite cosmology. Kepler did not find the discovery of new *moons* or satellites rotating about one of the planets in the solar system disagreeable, but the discovery of new planets revolving around one of the fixed stars would threaten his cosmology and seemingly support the ideas of Bruno and Wackher von Wackhenfeltz, Kepler's friend and confidante, and enthusiastic supporter of Bruno's vision. If Bruno were right – if the solar system were no longer equidistant from the fixed stars; if the universe were centerless and infinite – then it would be necessary to discard the vision of a universe created for man and man as the lord of all creation.

Kepler was in no way prepared to discard these beliefs. The opening pages of the *Dissertatio* are extraordinary. With his characteristic sincerity, Kepler summed up the situation and his opinions *after* the announcement of Galileo's discovery of new stars, but *before* knowing which stars he meant. Even before they had actually read the *Sidereus nuncius*, Kepler and his friend, Wackher, had two different interpretations of it. Kepler believed it was possible that Galileo had seen four small moons rotating around one of the planets and Wackher instead was certain that new planets had been sighted circling a fixed star. Wackher had already suggested this possibility to Kepler, "drawing from the speculations of Cardinal di Cusa and Giordano Bruno." Upon reading Galileo's text, Kepler felt vindicated and relieved. He wrote, "If you [Galileo] had discovered planets rotating around one of the fixed stars, then I would have been consigned to the shackles and bondage of Bruno's innumerability, or better yet, exile into that infinity. For the time being I am however liberated of the great fear that I felt, upon learning of your book, at the prospect of hearing my old adversary cry out in triumph" (Kepler, 1937–1959: IV, 304).

Kepler remained firm in his belief that the Earth was a privileged place in the universe and the only one suitable for man to inhabit. The planetary system to which we belong, according to Kepler, was located "in the center of the universe and around its heart: the sun." Within that planetary system, the Earth was located at the center of the primary globes (Mars, Jupiter, Saturn were the outer planets and Venus, Mercury, and the sun the inner ones). One could make out Mercury from the Earth but it could not be seen from Jupiter or Saturn. The Earth was the "the seat of that rational being for whom the universe was created," and the place dedicated to the most important and noble of corporeal beings" (Kepler, 1937–59: VII, 279; IV, 308).

Whereas Bruno was excited by the infinity of the cosmos, Kepler seemed to feel it was the source of "a secret and hidden horror: we feel lost in an immense expanse which has no boundaries or center, and because of this we are denied a determined place" (Kepler, 1858–71: II, 688). Kepler countered the theory of the infinite universe with a very "strong" one of his own, the importance of which would be discovered only two centuries later. He claimed that Galileo believed that in addition to the fixed stars that had been known since antiquity, the heavens were populated by over 10,000 more stars. Kepler came up with this figure based on a gross approximation, but what was important to him was that "the greater the number and mass of stars, the more valid my argument against the infinity of the universe" (Kepler, 1972: 55). Even if out of only 1,000 stars none were bigger than $\frac{1}{60}$ of a degree or a minute (and those that had been measured were larger), then gathered together they would be equal to or greater than the diameter of the sun. And what of 10,000 stars then? If those suns are made of the same material as our sun, "why would all those suns massed together not outshine our sun?" (ibid.: 55). This argument is the historical source of the famous "paradox of the night sky" discussed by Edmund Halley in the 1720s, and by the German astronomer Heinrich Olber's argument exactly one century later.

Galileo, Descartes, and the Infinite Universe

Galileo does not mention Giordano Bruno in any of his published works or letters, for which Kepler scolds him. And Alexander Koyré has analytically documented (Koyré, 1957: 95–8) that Galileo did not engage in the debate over whether the universe was finite or infinite. Instead he remained undecided (though more inclined toward the latter), considering the problem unsolvable: it has not been proved, nor will it ever be, that "the stars of the firmament are arranged in a sphere," and no one knows or will ever know its shape or whether it even has one" (Galileo, 1890–1909: VI, 523, 518). In the *Dialogue* he states that "the fixed stars are many different suns" and that "we do not know where the center of the universe is, or even if it has one." Yet the text also contains a firm rejection of the infinity of space (ibid.: VII, 306). In 1639 he wrote to Fortunio Liceti that there were "subtle reasons" for both views, but "to my mind neither one nor the other is conclusive so that I remain in doubt about which of the two answers is true." One argument in particular inclined him more toward the theory of the infinite universe: incomprehensibility better relates to incomprehensible infinite than to finiteness which is not impossible to comprehend. But, this is one of those questions – like predestination or free will – "that is happily inexplicable to human reason" (ibid.: XVIII, 106).

The arguments in Galileo's *Letter to Liceti* do not lack in subtlety. He writes, "if I have doubts with regard to the question of whether the universe is finite or infinite, if I cannot decide, then it is more likely that the universe is infinite because if it were finite then I would not have so many doubts." Descartes, on the other hand, took another approach in the *Principles* (1644). "We should never enter into arguments about the infinite," he wrote, "because it would be absurd for us, being finite, to attempt to understand a thing and imagine it to be infinite." For the finite mind to study the question of the infinite is to reduce it to the finite. Only those who believe their mind is infinite bother to ask such questions as, is half of an infinite line also infinite, or whether an infinite number is even or odd. We should refuse to answer such questions: "we should not try to understand the infinite but only think that if we are unable to discover a limit of something then it should be regarded as indefinite" (Descartes, 1967: II, 39). For with regard to the extension of the world, as with a series of numbers, one can always "go on and on." "Let us call these things *indefinite* rather than *infinite* so as to reserve that word for God alone" (ibid.: I, 39–40).

The correspondence between Descartes and the English Neoplatonic philosopher Henry More (1614–87) is tinged with references to Bruno,

Lucretius, the Cabal, and Cartesian philosophy. In these letters, Descartes
further refines his distinction between the indefinite and the infinite. In
answer to More's objection that if extension were limited and the number
of vortices finite then centrifugal force would cause the entire Cartesian
world system to scatter into atoms and dust, Descartes declared that extension was indefinite. For it was impossible to imagine a place outside
of the extension of matter into which these particles could flee.

In a universe without boundaries or center, the notion of man's centrality in the universe became meaningless. Anthropocentrism is a manifestation of pride; it is at once the manifestation of the inability to fathom
the greatness of the Creator together with the pretense of imposing our
privileged point-of-view on all creation. To believe that God created all
things for our use alone is presumptuous: "It is hardly likely that all
things were created for our use and so God had no other intention in
creating them [. . .] there are an infinite number of things in the world
that now exist, or that have existed and now have completely ceased to
exist, which man has never seen or known, or possibly had use for"
(ibid.: II, 118). Descartes had already stated in 1641 that because we can
never know God's intentions, it would be foolish to claim that God, in
creating the universe, had no other goal but to exalt man and that he
only made the sun in order to provide man with light (Descartes, 1936–
63: V, 54). The smallness of the Earth in relation to the largeness of the
cosmos is only incredible to those who have an insufficient understanding of God and who maintain that "the Earth is the primary part of the
universe because it is the home of man, and all things were created for his
benefit (Descartes, 1967: II, 138).

Descartes remained unresolved on the question of whether there was
life on other planets and other intelligent beings in the universe, though
he did believe that the mystery of incarnation and God's other gifts to
man did not prevent "him from having granted an infinite number of
similar gifts to an infinite number of beings." He added that he would
"leave such questions open, preferring neither to deny nor affirm anything" (ibid.: II, 626–7). However, at the end of his life, and in a spirit of
debate over anthropocentrism, Descartes once again proposed the idea
of many inhabited worlds. Man commonly believes he is "special to God,"
and for this reason feels that everything was created for his benefit alone,
and that his Earth comes "before all else." But how do we know if God
has created something else in the stars, and peopled it with "different
creatures and beings, perhaps even men, or beings like men?" The power
of God might be manifested in the creation of an infinite number of beings: we should never be too sure that everything is in our power and for
our use, when perhaps countless other creatures, perhaps superior to us,
exist elsewhere" (ibid.: II, 696).

We are not Alone in the Universe

Kepler believed in the existence of a wall or "vault" (he also used the expression *cutis sive tunica*, skin or garment) that enclosed a vast cavity with the sun at its center. Tycho Brahe believed in a finite universe bounded by the sphere of the fixed stars. Galileo believed in the inevitability of uncertainty. The five revolutionary cosmological theses discussed above are not a part of the "rigorous" discourse of the great astronomers of the seventeenth century. Instead, these were firmly held cultural notions (later reflected in the cosmology) that drew from a unique blend of Democritean and Lucretian themes, Copernican doctrine, Neoplatonism, and Hermeticism.

Platonism and Hermeticism were a vital part of the philosophy of the infinite: from Nicholas of Cusa to Stellatus Palingenius, from Thomas Digges to Giordano Bruno and Henry More. William Gilbert, who was deeply indebted to Hermetic vitalism, wrote in *De magnete* (1600) that "the great and many lights" of the fixed stars were to be found not in a spherical surface, or in a vault, but at different and very great heights. The question of the infinity of the universe was intertwined with the debate over the plurality of inhabitable worlds; a debate with ancient roots, the main principles of which were carefully explained in the great early sixteenth-century encyclopedia by Giorgio Valla, *De expetendis et fugiendis rebus* (published posthumously in 1501). In 1567 Philip Melanchthon advanced a number of physical and theological arguments against the claim that life in other worlds existed. His ideas were taken up countless times, with greater and lesser polemical force, by both Protestants and Catholics. In 1634, Kepler published the *Somnium seu opus posthumum de astronomia lunari/The Dream*. It marked the transition from "fantasy" literature about the moon (inspired by Lucian and Ariosto) to "science fiction," and for three centuries (until Jules Verne and H. G. Wells) inspired countless tales about journeys to the moon. *The Dream* is filled with veiled autobiographical allusions and references to tragic experiences in the author's eventful life. Nor was it a brief foray into creative writing for Kepler. The original idea dated to an early essay from 1593, the story was written in 1609, and between 1622 and 1630 Kepler added many (and long) notes (Rosen, in Kepler 1967: 21). The description of the trip to the moon is a unique blend of fantasy and realism. The inhabitants of the moon are gigantic, "serpent-like" creatures with short life spans; they bask in the extreme heat of the sun and then retreat to cold caves and crevasses. Yet the physical description of the moon-world is not fantasy; it reflects the telescopic discoveries of Kepler's day (Nicolson, 1960: 45). Kepler writes, "To us who inhabit the Earth, our moon, when it is full and rising and climbing above distant houses, seems equal to the rim of a keg; when it mounts to mid-heaven, it hardly matches the width of the human face. But to the Subvolvans, their Volva in mid-heaven [. . .] looks a little less than 4 times longer in diameter than our moon does to us. Hence, if the disks are compared, their Volva is 15 times larger than our moon. The inhabitants of the moon know that our Earth, which is their Volva, revolves, but their Earth is immobile" (Kepler, 1972: 6–7, 34; 1967: 21).

Four years later, John Wilkins (1614–72) published one of the most important "popular science" books of the seventeenth century. The *Discovery of a New World, or a Discourse Tending to Prove that it is Probable there May be Another Habitable World in the Moon* (1638) circulated widely and was liberally plagiarized by Fontenelle. In defense of his argument, Wilkins reminded readers that Columbus's voyage had been met with disbelief, that popular opinion is dogmatic, that it is typical to scorn new ideas; and that for centuries scholars had refused to believe in the existence of the antipodes. He was clearly aware of the theological threat inherent in the theory of habitable worlds, for the claim had been considered heretical since antiquity. If all worlds are alike, God is not "provident" for no world is greater than another; if they are different, none can be called "world" or "universe" for it lacks universal perfection. It is significant that of the standard arguments used *against* Copernicanism and the theory of the plurality of worlds, Wilkins cited the classic "diabolicentric" theory and "the base nature of our Earth, which is composed of the filthiest and most vile matter in all the universe, and so must necessarily be at its center, given that this is the worst possible position and the farthest away from the pure, unchangeable bodies in the heavens" (Wilkins, 1638: 68).

Copernican doctrine was opposed *also* on the grounds that it placed man in a too-elevated position by transporting him to places not unlike the unchangeable and immortal heavens. Similarly the consideration of other habitable worlds brought up some unsettling questions: what is the meaning of the fall from grace and of redemption, of original sin and Christ's sacrifice, if the Earth, the stage upon which this drama unfolds, is but one of many worlds? And if many inhabited worlds exist, then did the Savior redeem them too? If even the heavens are corruptible, then how can they possibly be the throne of the Lord?

Wilkins cited passages on inhabited worlds from Campanella's *Apologia pro Galilaeo* (1622) as a reliable source. Between the late 1630s and the 1660s a number of "science fiction" works appeared that merged the theme of space travel with philosophical, moral, and astronomical questions. Two examples were *The Man in the Moon* (1638) by Francis Godwin and *Description of a New World* (1666) by Margaret Cavendish.

Cyrano de Bergerac's *Histoire comique des états et empires de la Lune* (1656) and Pierre Borel's *Discours nouveau prouvant que les astres sont des terres habitées* (1657) were published within a year of one another in France. Cyrano, the notorious freethinker, believed in an organic and animate universe. He made references to Campanella, Gassendi and la Mothe le Vayer, and combined ideas drawn from Neoplatonism and the Cabal, the atomism of Democritus and Epicurus, Averroism, and the new cosmology of Copernicus, Galileo, and Kepler. The fixed stars were like suns and this proved the world was infinite "because it is likely that the inhabitants of a fixed star discover other fixed stars above them which we are not able to see from here, and this is repeated on through infinity." Just as anyone who has ever been in a boat believes the shore to be moving, so man believes that the heavens move around the Earth. This error in perception has been compounded by "man's insufferable pride which allows him to believe that nature was created only for him, as if it were likely that the sun had been lit only to make his apricots ripen and his cauliflower grow." Borel believed that Galileo's discoveries proved not only the reality of the Copernican system but also the validity of the theory of habitable worlds. While his text (like Cyrano's more seductively written work) contained no original ideas, it brought together the different elements of a complexly woven discussion drawn from a variety of traditions. Borel dedicated his book to Kenelm Digby and concluded with a long passage from Stellatus Palingenius. Copernicus, Kepler, and Campanella are names that occur with great frequency, and while there is no mention of Bruno, his ideas and the Lucretian worldview (the text refers to *De rerum natura* a number of times) are constantly in the background. Yet the most beloved master of all was Montaigne, who like Socrates, taught us to reject certainty and to doubt.

The famous works of Bernard le Bovier de Fontenelle (1657–1757) and Christiaan Huygens (1629–95) represent but the outcome of a discussion that had been unfolding over almost two centuries. Fontenelle's *Entretiens sur la pluralité des mondes* (1686) was reprinted 31 times during his lifetime. Next to the Cartesian theory of vortices, it familiarized a large readership with the theories of the infinite universe and the plurality of inhabited worlds. Fontenelle believed there was other life in the universe and supported his ideas by pointing to recent microscopic discoveries. The Marchioness character in the book is bewildered at the thought of an infinite and infinitely populated universe, but her teacher reacts differently to the same idea: the infinite puts him at ease, "if the heavens were but an azure vault imbedded with stars, the universe would seem small to me and this would sadden me [. . .]. Now the universe has other splendors and nature, in creating it, has spared nothing."

Huygens' Conjectures

When the great Huygens died in 1695, the *Cosmotheoros sive de terris coelestibus earumque ornatu conjecturae* (1698) had not yet been published. It was his opinion that neither Cusa, Fontenelle, nor Bruno had looked seriously enough into the question of life on other planets. He believed that no obstacles lay in the path of learning about such faraway things and material abounded for a series of probable conjectures. This work should not be discouraged for two reasons: first, if human curiosity had known bounds, then we would not yet know the shape of the Earth or that America existed; second, the research of probable theories is the very essence of the science of physics (Huygens, 1888–1950: XXI, 683, 687, 689).

Whoever witnesses the dissection of a dog would not hesitate to state that similar organs exist in an ox or pig. By the same token, our knowledge of the Earth allows us to conjecture about other planets. Gravity surely does not exist solely on Earth. Why should plant and animal life only exist there? It is true that nature seeks variety and that variation is the expression of the Creator's existence, but it is also true that plants and animals in the Americas are structurally similar to European plants and animals. The differences in life forms on other planets vary according to their distance from the sun, "but they probably differ in matter rather than form" (ibid.: XXI, 699, 701, 703). The wonders of plant reproduction "cannot have been solely invented for our Earth." This is not to say that the inhabitants of other planets are like us, but they are surely similar structurally: they are probably rational beings with values like our own, and have eyes, hands, writing, societies, geometry, and music (ibid.: XXI, 707, 717, 719–51).

Before the invention of the telescope, the theory that the sun was one of the fixed stars seemed to contradict Copernican doctrine. Today, "all who embrace the Copernican worldview" agree that the stars are not to be found on the surface of a single sphere but "are strewn throughout the vastness of space and that the distance between the Earth or the sun and the stars closest to them is repeated between those stars and the next, and those and the next, and so on" (ibid.: XXI, 809).

There are some interesting elements present in Huygens's criticism of Kepler on exactly this subject. Kepler, wrote Huygens, had a completely different view. Even though he believed the stars were spread throughout the vast sky, he claimed that the sun was located at the center of a large space, above which stretched the starry heavens. If the situation were different, he felt, then only a few stars would be visible and they would vary in size. In fact, argued Kepler, given that the largest stars seem so small that we can barely measure them, and that stars that are two to three times farther away would necessarily appear two to three times smaller (assuming they are all of equal size), then eventually we would arrive at stars which defy observation and two things would result: we would see very few stars, and they would be different sizes. However, just the opposite is true: we see many stars, and they do not appear very different in size. Huygens was convinced that Kepler's reasoning was flawed: he had not taken into account that the very nature of fire and flame is to be seen at distances from which other objects cannot be seen. For example, looking down a city street we might be able to count twenty or more lanterns, even though each one is 100 feet beyond the previous, and even though the flame of the twentieth is seen under an angle of barely six seconds. There is nothing strange, then, about seeing one or two thousand stars with the naked eye and then seeing twenty times more with a telescope.

Yet it was Huygens' belief that Kepler's mistake was even more pro-found. He *wanted* (*cupiebat*) "to consider the sun as preeminent among the stars; unique because it had its own planetary system and was located at the center of the universe." This was a prerequisite for proving Kepler's "cosmographical mystery," according to which a planet's distance from the sun corresponded to the diameters of the spheres inscribed and cir-cumscribed by Euclid's polyhedrons. For this reason, "there could only be one group of planets in the universe that revolved around a sun itself considered to be the only one of its kind" (ibid.: XXI, 811). This mystery was deeply rooted in Pythagorean and Platonic philosophy: the propor-tions did not conform to reality and the arguments advanced in favor of a spherical outer surface for the universe were weak. Furthermore, Kepler's conclusion that the distance of the sun from the concave surface of the sphere of the fixed stars was 100,000 times the diameter of the Earth was based on the odd argument that the diameter of Saturn's orbit is to the lower edge of the sphere of the fixed stars as the diameter of the sun is to the orbit of Saturn (ibid.: XXI, 813).

Huygens compared Kepler's odd ideas to Bruno's theory that the sun and the stars were of like natures. "As the foremost philosophers of our time have agreed, the sun and the stars have but one nature, and as a result, we imagine a universe even grander than ever. Who can prevent us from thinking that each of the stars or suns has it own planets and moons? [. . .] If we imagine ourselves somewhere in the heavens, in a point as far from the sun as from the fixed stars, we would see no differ-ence between them" (ibid.).

To make a mental leap like that of Huygens, to imagine oneself at a point in the universe that is the same distance from the sun as the closest of the fixed stars, and to observe the sun and the Earth (now invisible to us) from that point, is to conduct a "mind experiment" which has little in common with Galileo's mind experiments in natural philosophy. The exercise demanded a detachment from the terrestrial or heliocentric perspective of viewing the universe; a sort of cosmological relativism that was becoming popular at the same time that cultural relativism was being introduced. Huygens's very words reveal this: "We need to place ourselves outside of the Earth and in a position to look at it from afar. Only then can we ask if it is possible that nature bestowed all of its glories on it alone. Only in this way can we better understand the Earth and study it. By the same token, he who travels widely is a better judge of his homeland than he who never leaves it" (ibid.: XXI, 689).

The End of Anthropocentrism

A "Lucretian" world view was slowly developing. For at least a century (until the time of Baron d'Holbach and later) it represented an alternative to deism and the worldview constructed by Newton and the Newtonians. This new vision of the cosmos did not celebrate a perfect and ordered universe in which the designs of infinite wisdom were revealed for man's enlightenment.

Pierre Borel wrote that men must try not to be like those peasants, who having never seen a city, continue to claim for the rest of their lives that nothing could be greater or more beautiful than their own little village (Borel, 1657: 14, 32). The Earth is like a province or a village with respect to the universe, and he compared the situation to that of Europeans when they first learned about the discovery of new faraway lands and peoples.

The endless debate about the infinite nature of the universe and the multiplicity and inhabitability of other worlds made a great cultural contribution. It challenged every anthropocentric and "terrestrial" notion of the universe and rendered meaningless the traditional humanistic belief in the nobility and dignity of man. This discourse had to be reframed and contextualized in order to have more than just a rhetorical or literary function. The new image of nature and man's place in it, like the concept of the infinite universe, was used in different ways: it formed the basis of Pascal's profound religiosity as well as providing a foundation for the determinism of the great eighteenth-century materialists.

Bruno and Wilkins, Borel and Burnet, Cyrano and Fontenelle were the main characters in the complex drama that began with a closed world and led to an infinite universe. They liberally adopted the stunning astronomical discoveries of the great seventeenth-century astronomers to support their cosmological visions. Naturally their extrapolations were not always so legitimate or prudent. They argued from analogy. Yet even their "fantasies" contributed significantly to changing the course of the history of ideas and even the history of science. Kepler's *Somnium* and Huygens's *Cosmotheoros* show that the great scientists of the age were not indifferent to these "fantasies." It seems that imagination and cosmology are not so antithetical. After all, didn't one of the most important present-day cosmologists, Fred Hoyle, also write *The Black Cloud*?

9
Mechanical Philosophy

The Need for Imagination

The period between Copernicus and Newton was one of both the *macro-sciences* and *micro-sciences*. The former, such as planetary astronomy and terrestrial mechanics, deal with properties and processes which can be more or less directly observed and measured. The latter, such as the theories of optics, magnetism, capillarity, heat, and chemical change, postulated micro-entities, were regarded as unobservable on principle (Laudan, 1981: 21–2). Galileo, Descartes, Boyle, Gassendi, Hooke, Huygens, and Newton all discussed entities which were radically different from everyday macroscopic ones. In this context, the use of metaphor played an important role.

In mechanical philosophy, the physical world is represented by particles of matter in motion and can be interpreted by the laws of motion determined by statics and dynamics. A study of the physical world is therefore refined to the simplest terms and involves the *abstraction* of sensory or qualitative elements. As a result, a scientific *fact* is a material element that has been determined on the basis of strictly theoretical criteria. Experience came to be interpreted on the basis of pre-established theories. Natural phenomena such as air resistance, friction, the different behaviors of individual bodies, and the qualitative features of the physical world were now considered irrelevant to the discourse of natural philosophy or viewed as *disturbing circumstances* which were not to be (or should not be) taken into account in an explanation of the physical world. Mechanical philosophers were no longer fascinated by the peculiarities or the immediacy of familiar, everyday natural occurrences, nor by the "strange and curious" things which had been irresistible to Renaissance naturalists and magicians.

Given that an object's name and the object itself are not similar in any way, Descartes asked why nature could not have established a sign that would give us the sensation of light without having to resemble that sensation in the least? For example, philosophers had established that a sound is a vibration of the air, but our sense of hearing makes us think of sounds and not of air movements. By the same token, the sense of touch creates an awareness of an idea which is in no way similar to the objects that produce it. For instance, tickling is a concept that is not at all like a feather brushed against the lips. It is precisely this *non-similarity* that forces us to invent or imagine a model. What appears as "light" is actually a rapid movement transmitted to our eyes through the air and other transparent bodies. The analogy of a blind man who *sees* with his cane allows us to create and comprehend a model.

In the *Dioptrique*, Descartes uses analogies to support his mechanical theories: the blind man who "sees" with his cane illustrates the instantaneous transmission of light; wine that flows out of a vat because it is under pressure from all directions explains propagation; and a ball thrown off-course because it collides with another body describes the phenomena of refraction and reflection (Descartes, 1897–1913: XI, 84, 86, 89).

Science, by necessity, had to move from the *observable* to the *unobservable*, and it was the task of human imagination to conceive of the latter as similar in some way to the former. Science *obliges man to use his imagination*. Gassendi made the point that when we observe an attraction or a union we imagine hooks and ropes, something which grasps and something which is grasped; and when we see separation or repulsion, we imagine something that stings and something that is stung. It follows that in order to explain phenomena that are beyond human sensory perception, men are forced to imagine small pointers, stingers and other similar instruments that, although unobservable and incomprehensible, cannot be assumed not to exist (Gassendi, 1649: II, 1, 6, 14).

Robert Hooke was one of the seventeenth-century scientists who actively took part in the discussion about the composition of matter. In *Micrographia* he writes that because our sensory organs do not permit us to see how nature actually works, we hope that one day the microscope will allow us to observe the true and indivisible structures of bodies. In the meantime, he believed man was forced to grope in the dark and by "similitudes and comparisons" imagine the true causes of things (Hooke, 1665: 114). Hooke's meaning was clear: the internal structure of all matter and living organisms defied detection by the human senses (Hooke, 1705: 165). Because of this limitation, it is necessary to draw analogies between the effects produced by hypothetical bodies and those produced by causes accessible to the senses. From an *analogy of effects* one moves to an *analogy of causes*.

Hooke was a scientist in the Baconian tradition. He employed a method based on similarities, comparisons, analogies, and the application of an analogy of effects to an analogy of causes. So, for example, he explained what happens to air in the process of combustion; used experiments with a pneumatic pump to study meteorological phenomena; used a capillary model to explain the rise of liquids in filters and the lymphatic circulation of plants; and employed the law of elasticity to explain geological phenomena, such as the formation of springs. Hooke believed that the results of his research on light could be extended to magnetism, rarefaction, and condensation.

Mechanisms and Mechanics

Even the term *mechanization* is an elastic one, not given to a single definition and destined to have several different and vague meanings. The Dutch historian E. J. Dijksterhuis, in his history of the mechanization of the world from the pre-Socratics to Newton, asked himself the following questions: does the term which he applies to the millennia of development in scientific knowledge refer to the meaning of the word *implement* or *machine* implied in the Greek word *mechané*, to a world-picture in which the entire universe is like a big clock made by the Great Clockmaker? Or do we use it to mean that all natural phenomena can be described and understood with the aid of the concepts and methods that belong to that branch of physics called *mechanics*, which in this case means the science of motion?

Like many other historians of science, Dijksterhuis prefers clear-cut answers. He well knew that in the course of the seventeenth century, mechanics, as part of the science of physics, had been largely emancipated from its practical origins, its early ties to machines, and the environment of artisans, engineers, workshops, and mechanics. With Galileo and Newton, mechanics developed into a branch of mathematical physics that deals with the laws of motion (dynamics) and with bodies at rest and forces in equilibrium (statics). The so-called "theory of machines" is only *one* of its many practical applications. Many philosophers and historians of science seem genuinely disappointed that history (including history of science) is filled with instances of misunderstandings and misinterpretations. Dijksterhuis suggests that if mechanics had shed its ancient tie to machines and had been called *kinetics* or the study of motion, and if we had only called it the mathematization rather than the mechanization of nature, then many of these misunderstandings and misinterpretations could have been avoided.

However, it is hardly productive to try and solve the problems of history in terms of misunderstandings or linguistic ambiguities. It is in fact striking how often the new worldview presented by the seventeenth-century advocates, and even opponents, of corpuscular or mechanical philosophy reveal both of the meanings identified by Dijksterhuis, either in combination or mixed together. "Mechanical philosophy" (which prior to Newton was nothing like the branch of physics we now call *mechanics*) was based on the following assumptions: (1) nature is not the manifestation of a living principle but is a system of matter in motion that follows laws; (2) the laws of nature are mathematically precise; (3) relatively few such laws suffice to explain the universe; and (4) the explanation of natural phenomena excludes all reference to *vital forces* or *final causes*. On this basis, any explanation of natural events requires the building of a mechanical model as a "substitute" for the actual phenomena being studied. The model as a reconstruction of events is all the more valid (corresponds to reality) the more quantitative – and so geometrized – it is.

The immediate world of everyday experience *is not real* and in any case is completely irrelevant to science. Particle matter and its motion (based on laws) are real. The real world is a tapestry of quantitative and measurable data, of space and motion and connections in space. Dimension, form, and the state of movement of particles (and for some even the impenetrability of matter) are the only properties which are both real and explanatory principles of reality. Bacon, Galileo, Descartes, Pascal, Hobbes, Gassendi and Mersenne all distinguish between the *objective* and *subjective* qualities of bodies. This idea was one of the fundamental theoretical premises of mechanics, and led the philosopher John Locke (1632–1704) to articulate the celebrated distinction between *primary* and *secondary* qualities. And the doctrine was used to interpret and explain secondary qualities as well. In the *Leviathan* (1651), Thomas Hobbes (1588–1679) wrote: "all the qualities known as sensible are, in the object determined by them, different types of motion of matter which influence our organs in different manners. In we, who are equally stimulated, they (the qualities) are nothing more than different movements because movement produces nothing but movement, but their outward appearance is in our imagination [. . .] So the senses, in every case, are nothing but an image caused by a stimulus, that is by the motion exerted by external things above our eyes, ears, and other such organs" (Hobbes, 1955: 48–50). Even secondary qualities appeared to be mechanized *ex parte obiecti* and the very phenomena of sensation were reduced to a mechanical model.

Even Kepler, who was so deeply attached to Hermetic ideas, specifically likened the universe to a machine. In response to those who believed that "spirits" were responsible for celestial motion, he rejected the analogy between the universe and an animate divine being in favor of describing the universe as a clock: the different movements that occur in the universe are caused by a simple active material force just as the movements in a clock are simply the result of the pendulum. Boyle, too, believed the universe was like a great moving machine: "even though we should grant the Aristotelians that the planets [. . .] are moved by angels or immaterial intelligences; yet to explain the stations, progressions, and retrogradations and other phenomena of the planets, we must have recourse [. . .] to theories wherein the motion, figure, situation and other mathematical and mechanical affections of bodies are mainly employed" (Boyle, 1772: IV, 76).

Hobbes asked "why may we not say that all *automata* (engines that move themselves by springs and wheels as does a watch) have an *artificial life*? For what is the heart but a spring; and the nerves but many strings; and the joints but wheels?" (Hobbes, 1950: 3). Our body's mechanisms, stated Marcello Malpighi (1628–94) in *De pulmonibus* (1689), are the basis of medicine for we speak of "cords, fibres, running fluids, cavities, canals, and filters and other such mechanisms" (Malpighi, 1944: 40). In *Treatise on Man* (published in 1644 though completed in 1633) Descartes makes this analogy: "We see that clocks, artificial fountains, mills and other such machines, although only man-made, have the power to move of their own accord in many different ways [. . .] Indeed, one may compare the nerves of the [animal] machine I am describing with the pipes in the works of these fountains, its muscles and its tendons with the various devices and springs which serve to set them in motion" (Descartes, 1897–1913: XI, 120, 130–1; 1985: I, 99, 100).

The references to clocks, mills, fountains, and hydraulic works are constant and persistent. In "mechanical philosophy" the references to mechanics as a branch of physics and mechanics as the functioning of machines are closely connected. For centuries the notion prevailed that the universe had been made not only *for* man but *like* him. The doctrine that microcosm and macrocosm are alike had produced an anthropomorphic view of nature. The science of mechanics, however, did away with any attempt to study nature from the anthropomorphic perspective. Those that adhered to mechanical philosophy believed its distinctive method to be so powerful that it could be applied to every aspect of reality: not only to nature but to the world of living things, not only to celestial movement and falling bodies but also to the sphere of human perception and emotion. Mechanization also concerned itself with physiology and psychology. Theories of perception, for example, seemed based on the hypothesis that particles, by means of invisible pores, penetrate the sensory organs and produce movements that are transmitted to the nerves of the brain.

Mechanization was not just a method; it confirmed the existence of scientific laws and rejected explanations invoking spirits or "vital forces" as un-scientific. It developed – and this was recognized at the time – into a true philosophy. Mechanical philosophy subsequently proposed its own "image of science." It defined what science *was* and what it *should be*. With the exception of theology, there was no area of knowledge that could escape the principles of mechanical philosophy. In this vein, Thomas Hobbes even classified politics under the heading of mechanical philosophy.

The Natural and the Artificial: To Know and to Make

Mechanical philosophy looked to machines as the models best able to provide explanations, and the machine could either be one that actually existed or one that could plausibly exist. Since each element (or "part") of a machine performs a specific function and each "part" is equally necessary for the operation of the machine, there can be no hierarchy of parts in the world-machine; no phenomena are more or less *noble* than others. The image of the universe as a giant clock overthrew the traditional image of the world as a sort of pyramid, with the basest things at the bottom and the most noble at the top, closest to God.

Understanding reality required knowing how the machines internal to the greater world-machine worked. Pierre Gassendi (1592–1655), a priest and professor of astronomy and mathematics at Digne (Aix), authored some subtle rebuttals to Descartes' *Meditations*. He opposed the Cartesian "full" and vacuumless universe and claimed instead that the universe was composed of indivisible particles that moved in a vacuum. In the *Syntagma philosophicum* (1658), he drew a clear analogy between natural objects and machines, or artificial objects: "Let us investigate nature the same way we investigate the things we ourselves create [. . .] Let us make use of anatomy, chemistry and other such aids in order to understand, to the best of our ability, bodies and by deconstructing them, know what they are made of and according to what principles, and see if others could have been or may be constructed" (Gassendi, 1658–75: I, 122b–123a).

Gassendi violently opposed Aristotelianism and the occult, and was a harsh critic of the Cartesians as well. He had libertine ideas and postulated metaphysical skepticism as the basis for the conscious acceptance of the limited, temporary and "phenomenological" nature of scientific knowledge. Only God could know the essence of things. Man could only know those phenomena which he could represent with models or those artificial products (machines) which he had made with his own hands.

This assertion implied a substantial *non-diversity* between the products of art and nature and implied, moreover, the rejection of the tradi-

tional definition of art as *imitatio naturae*. If art were simply the imitation of nature, it could never achieve the perfection of nature. Art was
seen as an attempt to copy nature in its movements, and for this reason,
medieval texts had often suggested that the mechanical arts were forgerers
or *adulterinae*.

Mechanical philosophy also brought about a crisis in the relationship
between art and nature. Francis Bacon criticized the Aristotelian theory
of types according to which a product of nature (a tree) is defined as
having a *primary form* while the product of art (a table made from the
tree) has only a *secondary form*. According to Bacon in *De augmentis*,
this doctrine "has introduced a premature despair into human endeavor;
men should instead convince themselves that artificial things do not differ from natural ones in form or essence, but only in efficient cause"
(Bacon, 1887–92: I, 496). The ancients had said that the lightning bolt
could not be imitated but in fact modern artillery did just that. Art is not
simia naturae (the ape of nature) and does not "kneel before nature," as
medieval tradition would have it. Even Descartes agreed with this: "For
I do not recognize any difference between artifacts and natural bodies"
with the exception that the mechanisms of the man-made machine are
"large enough to be easily perceivable by the senses" while the "pipes
and springs which make up natural objects are so minute that they completely elude our senses" (Descartes, 1897–1913: XI, 21; 1985: I, 288).

Knowledge of final causes and essences was reserved for God, as the
creator or builder of the world-machine, and denied to man. The knowledge of how to make something or the link between knowing and constructing (or reconstructing) applies not only to man but also to God.
God understands the wondrous clock-universe because he *is* the clock
maker.

Man can only *truly* know that which is artificial or man-made.
Mersenne wrote that "it is difficult to discover truths in physics. Because
the objects of physics are in the realm of things created by God, it is not
surprising if we are not able to find their true causes [. . .] We can discover the truth only about those things which we make with our own
hands or mind" (Mersenne, 1636: 8). Though Hobbes the materialist
differed from Mersenne on so many points, he agreed on this one: "Geometry is provable because we are the ones who draw and describe its
lines and figures. And civil philosophy is provable because we construct
the state. Since we nevertheless do not understand the structure of natural bodies but seek it in their effects, there is no proof for the causes but
only a suggestion of what they might be" (Hobbes, 1839–45: II, 92–94).

This particular passage of Hobbes' has often been compared to
Giambattista Vico (1668–1744) and his famous principle of *verumfactum*. In *De nostri temporis studiorum ratione* (1709) Vico wrote that

"we prove the propositions of geometry because we do them, if we could prove those of physics we would do them." Similarly arithmetic and geometry, he wrote in *De antiquissima* (1710), are part of man's truth because their certainty is demonstrated by doing them. In *New Science* (1725 and 1744), Vico considered history an object of the new science precisely because it had been entirely created and made by man: in this long and dark night there shines but one ray of light: that the world of noble nations was certainly made by men" (Vico, 1957: 781).

The concept that knowing and making or doing are identical gave rise, as we have seen, to a new science aware of insurmountable limits. However, the notion also significantly affected morality, politics, and history.

Animals, Men, and Machines

In Descartes' physiology (or psycho-physiology), as outlined in part five of the *Discourse on Method* and in the *Treatise on Man*, living things are no longer distinguished as fundamentally different from that which is mechanical. Animals are machines. The existence of a rational soul serves to draw a line of demarcation not between the machine and living organisms, but between living-machines and the specific functions of that particular (and unique in our world) machine known as man, who alone is capable of thought and speech. Once the model of the machine has been adopted, he believed these were the only two functions which either could not be explained or at least not explained entirely to his satisfaction.

A machine that looked and functioned like a monkey would need, for each specific action, a particular arrangement of its organs. In Descartes' view, a machine that had the many and diverse organs necessary to act in all the circumstances of life, in the way in which human reason allows us to act, was inconceivable. In many instances the machine might react even better than man, but in other instances it would inevitably fail. Reason, or the ability to adapt to one's surroundings, was not a characteristic that Descartes believed a machine could acquire. The same was true for language. Though it was certainly possible to build a machine that could utter a few words or react with words to specific external stimuli, a machine would never be capable of coordinating language in order to respond meaningfully to received language.

So the spirit of reason did not therefore derive from the power of matter, but was intentionally created by God. However, everything short of thought and speech (which is admittedly a great deal) could be interpreted according to the strictest laws of mechanics. From Descartes' perspective, animals were simply machines, and the physiological life of man could be explained by the machine metaphor and brought back to a mechanistic model. In the first place, there is a difference between voluntary processes and purely mechanical ones. Man's soul, located in the pineal gland near the base of the brain, controls the muscles that transform thought into words and action. Breathing, sneezing, yawning, coughing, peristaltic motion of the intestines, blinking, and swallowing are natural and ordinary movements which depend on the "flow of the spirits." These "spirits," like "a breeze or fine flame," flow rapidly through the fine tubes of the nerves, mechanically causing the muscles to contract. The very force of the animal spirits that flow from the brain to the nerves explains this type of movement: for example, the foot's reaction when it is burned (the foot retracts, one cries out in pain and looks in the direction of the heat source), or the case of just decapitated heads that continue to twitch on the ground.

Descartes likened such actions to the movements of a clock or a mill, and he relied on a more complicated mechanism – the elaborate system of fountains in the royal gardens (like a seventeenth-century Disneyland) – for his metaphor of voluntary physiological processes. The simple flow of water activated a series of machines, played certain instruments, and even enunciated certain words. The nerves of the animal machine were like the pipes of the fountain, and the muscles and tendons like the springs and different engines that move it. The animal spirits were like the water that activates the fountain, its heart was the source of the water and the brain cavity its reservoir. The external objects that stimulated the sensory organs were like the people who stroll into one of the grottos and unwittingly become the cause of some of its movements. As they approach a bathing Diana (which they have caused to appear by stepping on a certain tile), the visitors cause a Neptune to suddenly advance and threaten them with his trident. The rational soul, which resides in the brain, performs the function of the fountain-keeper who must be stationed by the reservoir of water (where the pipes of the fountain return) in order to start, stop, or change their movements. In the post-cybernetics era, some have compared the Cartesian "fountain-keeper" to a self-regulating mechanism.

Descartes made a clear distinction between voluntary and involuntary physiological processes and seemed to understand the phenomenon which later would be dubbed (though to explain something very different) a "reflex action." He paved the way for the mechanical biology of the iatromechanics and the progressive substitution of the principles of traditional vitalism with the methods of chemistry and physics. However, there were dangerous implications in the theory of animals as machines; Jesuit priest Gabriel Daniel insisted in 1703 that all Cartesians as a result had to claim that human beings were likewise nothing but machines.

The Neapolitan astronomer and mathematician Giovanni Alfonso Borelli (1608–79) also compared automatons to self-moving animals, and he looked to geometry and mechanics as two steps in the path of "the marvelous science of the motion of living things." Shortly after he died, his greatest work, *On the Motion of Animals*, was published in Rome (1680–1). It was reminiscent of Harvey, sounded some Galilean themes from *Discourses*, and was Cartesian in approach. The movements of animals as they walk, run, jump, and lift, and the flight of birds and the swimming of fish, were studied from a geometric–mechanical perspective, as if they were systems of simple machines. In two parts, the book first discusses the external or apparent movements of the body and secondly the internal movements of muscles and organs, some of which were involuntary. The body is represented as a hydraulic mechanism in which animal spirits flow through the nerves like water. In most cases, the muscles labor at a considerable disadvantage: if the bones are a lever and the articulations of the joints the fulcrum, the muscle acts very close to the fulcrum while the weight (take the case of an extended arm holding an object) is close to the end of a lever which is ten or twenty times larger than the smaller lever that is represented by the muscle. The force exerted by the muscle is far greater than the weight it must lift.

Borelli worked from the Galilean–Cartesian premise that "through his works the Creator spoke in geometric language" (Borelli, 1680–1: I, 3r). In chapter two of *On the Motion of Animals* he wrote that "natural processes are simple and easy and follow the laws of mechanics, which are necessary laws." On the basis of these premises he rejected all chemical interpretations of physiological phenomena and described all bodily functions – circulation, heartbeat, respiration, and the functioning of the kidneys – in purely mechanical terms. He took chemical processes into consideration only in the case of the contraction and swelling of human muscles. Since the muscle fibers themselves are unable to contract on their own, they cannot independently lift weights by contracting, so the lifting "must occur due to an external force different from the material force of the machine that forcefully contracts it." In the face of such mysterious causes he "confesses ignorance" yet did not give up looking for the "probable causes" of natural processes. He felt it was necessary to go further and "hypothetically conjecture" about those things whose machinery was invisible. Borelli believed there had to be a happy medium between daring to observe no limits in philosophy and too hastily professing ignorance, although he, like Newton after him, argued that it was wrong to feign hypotheses: "non enim hypotheses fictas admittere debemus."

In *De venarum ostiolis* (1603), Girolamo Fabrici d'Acquapendente (1537–1619) compared the "membranes" in the veins to the stones placed in streams by mill builders so as to hold and accumulate the water necessary to operate the grinding mechanism. Similar "locks" or "breakwaters" could be found in the veins. Gabriel Harvey used a different machine – the pump – to illustrate the same idea, and so replaced the concept of the lock with that of the valve. Under the heading *méchanicien* in the great Enlightenment *Encyclopédie*, Diderot wrote that medicine had taken on a completely new aspect in the last 100 years, adopting a lexicon entirely different from the one that had long been in use.

Can Mechanists still be Christians?

The great seventeenth-century natural philosophers who had adopted and taught the science of mechanics were great admirers of the mechanical and corpuscular worldview constructed by Democritus and the ancient atomists, and by the Roman poet Lucretius. For the most part, however, they were careful to distance themselves from the heretical and atheistic implications of materialism, and eschewed philosophies that denied divine creation and attributed the origin of the universe to a *random* and fortuitous combination of atoms. In their view, the image of the universe as a machine implied a Maker or Builder, and the clock metaphor implied a Clockmaker. A careful and painstaking study of the great world-machine required reading the Book of Nature as well as the Book of Scripture, both of which reflect the glory of God.

Two philosophers in particular were frequently attacked and condemned in this regard: Thomas Hobbes (1588–1679) and Baruch Spinoza (1632–77). Hobbes applied mechanical philosophy to all of psychic life, conceived of human thought as an instinct a bit more complicated than animal instinct, and explained all natural processes and change in terms of movement. By attributing extension to God, Spinoza flaunted the traditional distinction between a material world and an immaterial God, and rejected God as a being with designs and intentions. Such plans were nothing but projections of exclusively human needs onto God. He declared body and soul inseparable and saw the universe as a purposeless, senseless, eternal machine which was the expression of a necessary and immanent causality.

In the late seventeenth and early eighteenth centuries, terms such as Hobbesian, Spinozan, atheist, and libertine were often synonymous with one another. The most radical theories of the libertine movement occurred in the widely read *Theophrastus redivivus*, written around 1666. And it was by that underground route that Renaissance Naturalism and heretical mystic-Hermetic themes were joined to the anti-Newtonian and anti-deistic philosophy of John Toland (1670–1722) and later to the works of the great eighteenth-century French materialists.

Although Pierre Gassendi, as we have seen, proposed that atoms were created by God, his views appeared dangerously close to libertine ones.

Marin Mersenne (1588–1648) openly attacked libertine positions in *L'impiété des déistes* (1624). He abandoned scholasticism to decisively take the side of new science, which he viewed as a defense against a great threat to Christian beliefs and values. This threat included a return to "mystical" themes, the spread of Hermeticism, and the holding of views rooted in Renaissance naturalism and in the doctrines of Pietro Pomponazzi (1462–1525), who refuted the existence of miracles and claimed that the three great Mediterranean religions had been founded for political reasons by three "impostors": Moses, Christ, and Mohammed.

Mersenne believed that natural magic, which allowed man to work "miracles," posed a greater threat to Christianity than the new mechanical philosophy did. The latter could in fact be reconciled with religion. He felt that the hypothetical and conjectural nature of scientific knowledge left all the necessary space for the religious dimension and Christian truth. Robert Boyle (1627–91) shared this concern and while he extolled the virtues of corpuscular and mechanical philosophy in *About the Excellency and Grounds of the Mechanical Hypothesis* (1655), he was at the same time concerned to establish two dividing lines: one between himself and the Epicureans and Lucretians who believed that nature and natural phenomena were produced by the random interaction of atoms in a void and the second between himself and the so-called "modern mechanists" (Cartesians) who believed that God had initially introduced a fixed quantity of motion into the total mass of matter, and that the various parts of matter, by virtue of their own motion, were able to independently form a system. Boyle's brand of corpuscular and mechanical philosophy was not to be confused with either Epicurianism or Cartesianism. Boyle's mechanical view separated the problem of the "origin of things" from that of "the subsequent course of nature." God had not limited himself to putting matter into motion, but guided the movements of individual particles of matter in such a way as to "contrive them into the world he designed." Once God had organized the universe and established "the rules of motion and that order amongst things corporeal which we are wont to call the Laws of Nature," it was possible to assert that worldly phenomena are "physically produced by the mechanical affections of the parts of matter, and what they operate upon one another according to mechanical laws" (Boyle, 1772: IV, 68–9, 76). The difference between *origin of things* and *subsequent course of nature* is very important: those who study the origins of the universe make the impious claim of deducing the world, and constructing theories and systems. Boyle felt that the Democritic–Epicureans and the Cartesians represented the atheistic and materialistic version of mechanical philosophy.

Indeed, what else had Descartes set out to do if not describe the creation of the universe in his treatise *The World* and present an alternate creation story to that of Genesis? True, Descartes had presented his description of the creation of the world as a "fable," and also claimed to be describing an imaginary universe. Yet he had curiously inverted the sense of his discussion in a number of places: by understanding how the fetus forms in the maternal womb and how plants grow from seed, we know more than if we simply knew the grown child or plant. In part three of the *Principles*, Descartes claimed that the same was true for the universe. Science could not only tell about the world but also about how it was formed. The contrast with Boyle on this point is radical. In chapter six of *The World*, Descartes wrote that "the laws of nature are sufficient to cause the parts of this chaos to disentangle themselves and arrange themselves in such good order that they will have the form of a quite perfect world." The present structures of the world, according to the Cartesian view, were the result of matter, the laws of matter, and time.

With respect to this doctrine and these answers, Isaac Newton's position was not far from Boyle's. From the start, Newton relied on the anti-Cartesian objections of Henry More (1614–87) and Pierre Gassendi: "If like Descartes we claim that extension is corporeal, are we not promoting atheism? For two reasons: because extension appears increate and eternal, and because in certain circumstances we may imagine it as existent and at the same time imagine the non-existence of God." Newton found the Cartesian dualism of mind and body incomprehensible: "for if mind is not extended in space, it is nowhere and does not exist" (Newton, 1962: 82, 109).

Newton's detachment of his philosophy from the potentially atheistic and materialistic outcomes of Cartesianism took a different form but was nevertheless a dominant theme, as he makes perfectly clear in *Query 31* of the *Opticks* (appended to the 1717 edition) and the *General Scholium*: "blind fate" could never make all the planets move one and the same way in concentric orbits, and the wonderful uniformity of the planetary system is the effect of "intentional choice." The planets keep moving in their orbits because of the laws of gravity, but the "original and regular position of these orbits cannot be attributed to the laws: the admirable location of the sun, the planets, and the comets can only be the work of an omnipotent and intelligent Being." The distinction that Boyle had made between the origin of things and the regular course of nature was used in this context. If it were true that "solid particles had been joined together in different ways at the creation on the advice of an intelligent Agent," and were arranged by their designer, then "it is unphilosophical to seek for any other explanation for the origin of the world, or to pretend that it might arise out of Chaos by the mere laws of nature" (Newton, 1953: 402–4). Natural laws only began to operate after the creation of the world. Newton's science was an exhaustive description of the universe as it was; from Moses' account of the creation of the world to the predicted Apocalypse. Newton and the Newtonians never accepted the idea that the world could have been *produced* by the laws of mechanics.

Leibniz and the Critique of Mechanics

Leibniz, too, believed that Cartesian philosophy, as the premise for all mechanical philosophy, was exceedingly dangerous. Descartes had written in the *Principles* that by means of the laws of nature "matter must successively assume all the forms of which it is capable; and, if we consider these forms in order, we will eventually be able to arrive at the form which characterizes the universe in its present state" (Descartes, 1985: 258). According to Leibniz, if matter can assume all possible forms then it follows that any absurd, bizarre, or unjust thing we are able to imagine has happened or may one day happen. So, as Spinoza would have it, justice, goodness, and order become simply concepts that are relative to man. If all is possible, and all that is possible is in the past, present, and future (so, too, claimed Hobbes), then there is no such thing as Providence. The Cartesian claim that matter successively assumes all possible forms is tantamount to destroying the wisdom and justice of God. Descartes' God, concludes Leibniz, "makes all things that are possible and proceeds, according to a necessary and fatal order, through all possible combinations: to this end matter was the only necessity, and Descartes' God no more than that necessity" (Leibniz, 1875–90: IV, 283, 341, 344, 399).

Leibniz viewed Cartesianism as materialism. In an autobiographical letter dated 1714, Leibniz described his first encounter with modern philosophers: "I remember being fifteen years old, and walking alone one day in the woods near Leipzig, debating with myself the theory of substantial forms. Mechanics ultimately won out and that spurred me towards mathematics. [...] Yet while studying the fundamentals of mechanics and the laws of motion, I returned to metaphysics and the doctrine of entelechy" (Leibniz, 1875–90: III, 606). This return to metaphysics proved extraordinarily significant in the progress of mathematics, physics, and biology. Next to Cartesianism and Newtonianism, the metaphysics of Leibniz was one of the great influences on science during the eighteenth century and even later.

The mechanical view, according to Leibniz, was a partial position that required integration within a broader perspective. While useful in physics, it was utterly useless in metaphysics. Studying the structure of the universe could not be separated from a study of God's "intentions." To discuss a construction is to actually penetrate the intentions of its architect; to explain how a machine works, it is necessary to "ask what its purpose is and demonstrate how all of the pieces serve that end." The modern philosophers were "too materialistic" because they limited themselves to dealing with the figures and motions of matter. Physics does not have to limit itself to investigating the nature of things without asking why they are as they are. Final causes did not simply serve for the appreciation of divine wisdom but "to understand things and use them" (Leibniz, 1875–90: IV, 339).

Leibniz criticized the very foundations of mechanics: the identification of matter as extension; the corpuscular quality of matter and it divisibility into indivisible atoms; the passivity of matter; and the distinction between the material and rational worlds.

Extension, which is geometrical, homogeneous, and uniform, does not explain motion or the resistance of bodies to motion. That resistance is in no way derived from extension. In 1686, Leibniz published the controversial "A Brief Demonstration of a Memorable Error by Descartes." The "memorable error" was Descartes' principle of the conservation of the quantity of motion (the product of mass and the velocity of a body). What is actually conserved is living force (*vis viva*) – later called kinetic energy – which is equivalent to the product of mass and the square of its velocity. Descartes and the Cartesians had confused quantity of motion with force, and the error was based on the use of simple machines as a model. Leibniz drew a clear line of demarcation between statics and dynamics (Westfall, 1971a).

Living or active force was much more to Leibniz than just a number or a mathematical quantity. It was a metaphysical reality whose characteristics not only contradicted the assumptions underlying mechanics but necessitated their reversal. Matter and motion, according to Leibniz, were the perceptible manifestations of a metaphysical reality. The active pole of this reality is *conatus* (a Hobbesian term) or energy or *vis viva* which appears phenomenologically as motion. The passive pole is the primary matter that appears phenomenologically as inertia, impenetrability, or resistance to impact of matter. Physical bodies or composed substances are the phenomenological products of metaphysical points or centers of force or simple, individual substances (directly created by God), which Leibniz gave the Pythagorean or Brunian name of *monad*. A monad could not be attained by simply subdividing matter: deprived of spatiality and shape, monads are self-contained and reciprocally independent entities ("they have no windows"). Each monad was endowed with symbolic force with regard to the rest of the universe and with the tendency to pass from one state to another. Monads are analogous to the human soul.

The theory of metaphysical points or centers of force re-established the link between the material and the spiritual, and once again questioned the deeply rooted Cartesian and atomistic notion of the qualitative difference between *res extensa* and *res cogitans*.

Leibniz rejected the idea of void space and forces at a distance (on this point agreeing with Descartes and radically disagreeing with Newton). In a series of letters (1715–16) with the Newtonian Dr. Samuel Clarke, he refuted absolute space: time and space being neither substances nor absolute beings, they are but the order of co-existences and the order of succession – they are "relative." In a letter to Father Honoré Fabri (1607–ca.1688), Leibniz clarified his position with respect to other schools and traditions: "Cartesians identify the essence of a body as its extension. Even though I reject void space (agreeing with Aristotle and Descartes and disagreeing with Democrites and Gassendi), I nevertheless believe there is something passive in bodies, that is to say that bodies resist penetration. In this regard, I agree with Democritus and Aristotle but not with Gassendi and Descartes" (Leibniz, 1849–63: VI, 98–100).

According to Leibniz, physics could not be reduced to mechanics and mechanics was not the same as kinematics (as it was for Descartes and Huygens). The model for physics was not a simple balance in a state of equilibrium, both forces appearing equal. Force equaled quantity of motion only in static situations (Westfall, 1971b: 136). Leibniz coined the term "dynamics" to describe a mechanics centered around the concept of force, and he used this term in *Essay de dynamique* (1692) and *Specimen dynamicum* (1695): "the notion of virtue or energy, *Kraft* in German and *force* in French, which I have explained by the science of dynamics, adds a great deal to our understanding of the essence of substances" (Leibniz, 1875–90: IV, 469).

The terms substance and activity overlap one another: substance is activity and where there is activity there is substance. Not all that exists is alive, but life is to be found everywhere. Leibniz found both evidence and stimuli for his system in the biology of his day. Microscopic discoveries, for example, appeared to support his vision of matter as an infinite aggregate of monads, where every fragment of matter is like a pond full of fish, and every part of that fragment is in turn like another pond. In the *Nouveaux essais sur l'entendement humain* (1703), which contain the famous attack on Locke's empiricism and a defense of virtual innatism, Leibniz predicted the ever growing use of the microscope in light of the ongoing identification of similarities between living things. Generation as development and growth (but the entire universe was, from his perspective, the unfolding of implicit possibilities present before creation and "programmed" like in an embryo) placed Leibniz squarely among the preformationists.

The harmony that reigns in the real world, chosen by God as the "best" among all possible worlds (a world is the sum total of all the possibilities that can coexist without contradiction), bars gaps, discontinuities, and contraposition from occurring in nature. Nature obeys the principles of continuity and plenitude: all the substances that have been created form a series in which every possible quantitative variation is possible. There is no room in the universe for two exactly identical entities that do not differ internally (the principle of indiscernibles). Contrary to Cartesian doctrine, God does not establish eternal truths. His actions are not arbitrary and He respects the principles of non-contradiction and increate logic.

Nothing exists or occurs without there being a reason for its existence or occurrence in a particular way and not another. Truths of fact are based on the principle of sufficient reason according to which nothing in the universe happens randomly or without cause. Truths of reason are supported by the principle of contradiction and for every true proposition the predicate must agree with the subject. Truth is not based on the Cartesian intuition of evidence but depends on the form of the argument. Essence or possible entities are governed by logical necessity, while existences or the actual entities that make up the world are subject to God's will and the governing principle of the best choice.

From God's perspective, truths of fact and truths of reason are the same. From man's perspective, in order to understand the real world, the deductions and rational explanations typical of formal sciences have to coexist and mesh with the study of why a particular phenomenon occurs in a particular way. The study of the natural world is not solely one of deductions; it is not just mathematics but experimentalism as well. The relationship between individual phenomena is of a mechanical nature but is based on a teleological order. Leibniz considered materialism and Spinozism as the illegitimate sons of the new science of nature.

10
Chemical Philosophy

Chemistry and its Forefathers

Astronomy and chemistry simply cannot be considered on the same level in a single, general discussion of the Scientific Revolution. Sixteenth-century astronomy was already a highly organized, mathematically sophisticated, theoretical science. Sixteenth-century chemistry, on the other hand, had no organized structure whatsoever, no theories of changes and reactions, and no clearly defined tradition. Like geology and magnetism, chemistry *became* a science between the seventeenth and eighteenth centuries. Unlike mathematics, mechanics, and astronomy, it was itself a *product* of the Scientific Revolution. Modern chemists do not descend from a long line of great and noble scientists dating to Antiquity and the Renaissance. There is no figure like a Euclid, Archimedes, or Ptolemy in the history of chemistry. Instead, modern chemists find themselves in the somewhat disconcerting company of alchemists, druggists, iatrochemists, sorcerers, astrologers, and other sundry figures.

The closest thing to an actual "chemist" (that is, closer to a modern chemist than an alchemist or a follower of Renaissance naturalism) appeared around the middle of the seventeenth century. However, with very few exceptions, that figure was not recognized as such, and was completely outside of the academic mainstream. He worked as a druggist or doctor, or in an Academy of mineralogy or metallurgy, or a botanical garden. Through his art, the doctor–chemist and the druggist–chemist succeeded in producing substances identical to those found in nature, and his practice was not necessarily divorced from a Paracelsian or occult context.

Without a doubt, *chemical philosophy* was rooted in the occult and its theoretical matrix was to be found in the grandiose writings of the intriguing (then and still now) Swiss Phillip Aureolus Theophrast Bombast von Hohenheim, better known as Paracelsus (ca. 1493–1541). Yet *chemical philosophy* occupies an important place in the scientific culture of the seventeenth century. Many of Descartes' or Campanella's contemporaries believed it to be as revolutionary and groundbreaking as the new mechanical philosophy. It effectively destroyed traditional Galenic teachings, profoundly transformed the practice of medicine, and changed university instructional methods. During the seventeenth century, Hermetic philosophy and Paracelsism were by no means fringe phenomena limited to small intellectual groups. Chemical philosophy and Paracelsian doctrines inspired a debate in Europe that was as far-reaching and intense as that aroused by Copernicus and the new astronomy. In fact, during Newton's formative years, the period of Puritan revolt (1650–70), Paracelsian influence was at its peak (Webster, 1982).

The Hermetic–Paracelsian tradition had very little impact on physics and astronomy. Yet to the empiricists and those who manipulated substances, it provided a cohesive theory which became the foundation for the study of substances and laboratory practice.

Paracelsus

Paracelsus led an extremely eventful life. He wandered throughout Europe rousing debate, controversy, and bitter argument everywhere he went. He burned the books of Galen and Avicenna on St. John's Eve 1527 in a student's bonfire at Basel. A man of violent actions, he attracted as many disciples as enemies. He believed that magic was a "great secret wisdom just as reason is a great public folly." He violently attacked those theologians who unjustly defined magic as sorcery without attempting to understand it, and even more violently attacked traditional medical practitioners and the university methods used to train them. He was himself an exceptional figure: the adjective "bombastic" derives from his name. According to Paracelsus, the four "pillars" of the new medicine were: *philosophy*, as knowledge of the invisible nature of things; *astrology*, or how the stars affect the human body; *alchemy*, or the preparation of remedies that can restore imbalances in the body that result from disease; and *ethics*, or the virtue and honesty of the physician. The close relationship between chemistry and medicine gave rise to a new discipline: *iatrochemistry*, or medical chemistry. Alchemy was predominantly a tool used in the distillation and analysis of minerals for the preparation of effective treatments.

Paracelsus believed that medicine should not deal solely with the human body: "One must understand that medicine is grounded in the stars, and the stars are the means for healing [. . .] a doctor should be trained in such a way that medicine operates through the heavens just as do prophecy and other celestial events" (Paracelsus, 1973: 136). The theory of a correspondence between macrocosm and microcosm was at the center of a group of ideas derived from astrology and the mystical–alchemical tradition crossed with themes common to Neoplatonic mysticism. The vital substances of objects are made up of invisible spirits or forces of nature. These spirits or *arcana* come from God who created things as *prime* and not *ultimate* matter: the world was a continuous chemical process of perfecting prime matter into ultimate matter. The Paracelsian "elements" were archetypes concealed in natural objects which gave them their characteristics and qualities. Those substances which could actually be treated and analyzed were but an approximation or a shell for the true spiritual elements. Prime matter, or the *Mysterium Magnum* or *Iliastrum*, was the mother or matrix of all things, and aqueous in nature. The three other traditional elements – fire, earth, and air – were also *matrices*. Plants, minerals, metals, and animals were the *fruit* of the four elements. In *Archidoxis* (written about 1525 and published posthumously in 1569) and *Liber de mineralibus*, the reader finds both the Paracelsian theory of the elements as the matrices of bodies and the theory of principles: Salt, Sulfur, and Mercury. This *tria prima* also consisted of spiritual substances and corresponded to the Body, Soul, and the Spirit. Salt makes bodies solid, Mercury makes them fluid, and Sulfur makes them inflammable. The three principles manifest themselves differently in different bodies and as many different types of sulfurs, mercuries, and salts exist as do species in nature: "One type of Sulfur is found in gold, another in silver, yet another in lead, and so on. There is another type of Sulfur in stone, lime, springwater, and salts. Not only are there many different Sulfurs, but many different Salts as well. One type is found in gems, yet another in metals, stone, salts, vitriol, and alum. The same is also true for Mercury" (Paracelsus, 1922–33: III, 43–4).

Chemistry was the key to unlocking the structure of the world, and creation was a divine chemical "separation." First the four elements were separated from one another, then Fire was separated from the Firmament, Air from the spirits, Water from marine flora, and Earth from wood, stone, land flora, and animals until individual objects and creatures remained. In *Philosophia ad Athenienses* (published in 1564) the process of creation as a whole was discussed in terms of alchemy.

Paracelsians

Paracelsism at the end of the sixteenth century was debated in a variety of works: *Idea medicinae philosophicae* by Petrus Severinus (Sørensen), published in 1571; the *Compendium* (1567) by Jacques Gohory (Leo Suavius, 1520–76), a lawyer in Paris Parliament and the French translator of Machiavelli; and in *Clavis totius philosophiae chymicae* (1567) by Gérard Dorn (?–1584). The *Basilica chymica* by Oswald Croll (1560–ca. 1609) was published the same year its author died, and by midcentury had been reprinted 18 times in both the original Latin and all major European languages. But the best-known work was that of Robert Fludd (1574–1637), whose writings from the period 1617–21 were discussed by Kepler, Mersenne, and Gassendi. In *Utriusque cosmi historia* (1617–18), a mystical–alchemical account of the creation became the basis of a *philosophia mosaica* in which Darkness, Light, and Water in Genesis were seen as the roots of the ancient doctrine of the four elements. Fludd was profoundly influenced by both Rosicrucians and Pythagorean number mysticism.

Paracelsus was the first to introduce the medicinal use of mineral substances to the practice of medicine. Chemistry, or the spagyric art, became one of the cornerstones of medicine. According to Joseph Duchesne (Quercetanus, ca.1544–1609), chemistry "teaches the composition, separation, preparation, alteration, and finally the emanation of all mixed bodies [. . .], it demonstrates the distillation process through the use of seven operations [. . .] to perfect all transmutations, by which we mean when a thing loses its extrinsic form and is so altered that it is no longer like its original form, but changes form and assumes another essence and color, and ultimately its very nature is changed and it takes on properties different from its original ones [. . .]. The seven Spagyric principles are: Calcination, Digestion, Fermentation, Distillation, Circulation, Sublimation, Fixation" (Quercetanus, 1684: 7).

The Belgian doctor Jean-Baptiste van Helmont (1579–1644) was another figure who constructed a complex chemical cosmology based on a "chemical" interpretation of Genesis. After Mersenne's *Questionnes celeberrimae in Genesim* of 1623 – in which he accused magic of being anti-Christian – alchemy and Paracelsian theories appeared even more threatening than previously. Van Helmont was questioned by the tribunal of Malines–Brussels on 24 propositions contained in his texts. He pleaded guilty and was sentenced by the Church in 1627 and again in 1630 after the department of theology at Louvain University and the Lyon College of Physicians brought new accusations against him for superstition and witchcraft. Arrested in March 1634, his books and manuscripts were impounded and he was transferred to a Franciscan convent in Brussels. This time he denied wrongdoing but was sentenced to house arrest for two years. In 1642 he was given permission to publish again. *Ortus medicinae* – 1,000 pages of his collected writings – appeared in 1648, four years after his death. This was one of the most popular scientific publications of the seventeenth century. By 1707, there had been seven editions in Latin, and it had been translated into English, French, German, and an abridged version in Flemish.

Van Helmont's animistic view of nature was based on the principle of movement. While he found the notion of parallelism between macrocosm and microcosm "poetic and metaphorical . . . [it was] unnatural and untrue." The only two elements in nature were air and water. Fire was not an element but a tool that could be used to modify the structure of bodies. The Paracelsian *tria prima* could be obtained by using fire to break down mixed substances. This notion of fire, according to which not only could previously combined substances be broken down but new classes of substances created, influenced Robert Boyle's theory of chemical elements (Abbri, 1980: 77). Some of Van Helmont's contributions to scientific development included: his interest in weight and quantification; his adoption of the theory of void space and attack on the *horror vacui*; his definition of a *gas* as something not *in* a body but the body *itself in* a different form from its original, and as a sign of imminent transmutation; and finally his explanation of digestion as the action of acid as an agent in the transformation of foods (Debus, 1977: 329–42).

The Iatrochemists

Chemistry in the seventeenth century was clearly already beginning to break free from the cosmological, biblical, and metaphysical background in which all discussions of principles, elements, substances, and the transformation of substances had been grounded. The path was anything but linear however, so there is a danger of isolating individual statements that suddenly sound "familiar' to us. The French translation of Jean Beguin's *Tyrocinium chimicum* (1610), part theory and part pharmacoepia, became widely known. Beguin writes that there are arts, like architecture, that bring life to their subject through the composition of parts and there are those, like chemistry, "that break down and open up their subject in order to see its innermost nature [. . .] and to obtain virtues either hidden, buried, or simply inefficient due to impurities, and confer upon them powers which are unimpeded" (Beguin, 1665: 27).

The ability to obtain secret powers had obvious practical applications, and nowhere was this more evident than in the work of the greatest seventeenth-century analytical chemist, the autodidact Rudolph Glauber (1604–68). Born in Karlstadt, he worked mainly in Holland. His *Furni novi philosophici oder Beschreibung einer neue erfunden Distillirkunst*, published between 1646 and 1650, was translated into Latin, French, and English. He described the new art of distillation (referred to in the title) which involved the production of hydrochloric acid, nitric acid, sulfuric acid and several derivative salts. When Glauber (by the action of sulfuric acid on sodium chloride) produced sodium sulfate (which together with magnesium sulfate became a fashionable drug), he named it after himself and kept the process a secret, earning considerable profits from it. He was so deeply indebted to Paracelsian metaphysics that he believed in the existence of a single original salt, and identified saltpeter (of great interest insofar as it was a component of gunpowder) as the universal salt. Glauber published an impressive six-part work on German prosperity between 1656 and 1661, entitled *Des Teutschlandts Wohlfahrt*. In it he claimed that chemical philosophy had the potential to right the disastrous wrongs produced by the Thirty Years War, and furthermore could secure Germany's position as "world monarch." Glauber wrote that "those who understand Fire and its uses will never suffer the ills of poverty. Those who do not understand it will never be able to look within the treasures of nature. Obviously, we Germans possess treasures of which we are not even aware and do not use to our advantage [. . .]. In fact, we devote more time to drinking and eating than we do to Art and the Sciences" (cf. Debus, 1977: 435).

Chemistry and Mechanical Philosophy

Robert Boyle (1627–91) has already appeared as a prominent figure in a previous chapter discussing the ideas, methods, and views of mechanical philosophy. In this context, Boyle believed chemistry was the science that could both *establish* mechanics and validate it. *The Sceptical Chymist* (1661), contrary to many textbooks, does not contain Boyle's theory of chemical elements. From his perspective, there were no qualitatively distinct chemical elements: matter was neither composed of the four Aristotelian elements nor Paracelsus' *tria prima* nor even the more recent *five principles* of French chemistry, but was a unitary, material reality composed of uniform particles which join together to create the materials that chemistry analyzes. His writings are very clear on this point: "I do not see why we must necessarily suppose there are primitive and simple bodies which nature is obliged to use in order to compose all the others. Nor do I see why we cannot imagine that nature is able to produce one mixed body from another through the different transformations of their minute particles without breaking down matter into any of the simple and homogeneous substances which some assume nature is broken into" (Boyle, 1900). He states just as clearly that "Salt, Sulfur, and Mercury are not the original and simple principles of bodies but rather primary clusters of corpuscles and simpler particles that seem endowed with the original or more radical or more universal affections – that is, size, shape and either movement or rest – of more simple bodies [. . .]. Our explanations are mechanical and simpler, and so must be considered more general and satisfactory" (Boyle, 1772: IV, 281).

Boyle believed in the transmutation of bodies as a necessary corollary to his corpuscular view of matter. The *tria prima* were clusters of particles produced by fire; and here he clearly borrowed from van Helmont. Boyle also studied combustion, calcination, and respiration. He rejected the idea of air as a simple and elementary body, and defined atmosphere as "a great receptacle or *rendez-vous* of heavenly and earthly emanations" (ibid.: IV, 85–6) made up of three types of particles: the first produced by vapors or dry fumes that rise from minerals, vegetables, and animals; the second type is finer and made from magnetic earthly fumes and countless particles emitted by the sun and other stars, and they produce what we call light; and the third class of particles "are not made elastic by external agents but are permanently elastic and may be defined by the term *permanent air*" (ibid.: V, 614–15). It was in this context that Boyle conducted his famous experiments on the elasticity of air and formulated what is known as Boyle's Law according to which there is a quantitative relationship between the pressure of a given amount of air and its volume.

Mechanics and Vitalism

Modern chemical theory recognizes the existence of elements or a precise number of substances that have been identified by a series of exact experiments. Because Boyle believed that chemistry could actually transform anything into any other thing, it has been pointed out that his work in chemistry appears to have been thwarted by his mechanical philosophy (Westfall, 1971: 79). However, it is also true that once chemists had adopted mechanical principles, chemistry was forever changed. Moreover, over the course of the entire seventeenth century, not only did the methods, principles, and underlying philosophy of chemists change, so did their *social status*. Their work was regarded with a respectability that it had not previously enjoyed.

Early in the eighteenth century, the physician Georg Stahl (1660–1734), a great German chemist, was entirely aware of how radically things had changed. "Chemistry," he wrote in 1723, "was for 200 years dominated exclusively by charlatans who produced an infinite number of victims [. . .]. Today there are some who have begun to seriously work in this science. That they are few in number should not be surprising. Naturally, all the impostors with their false promises of turning metals into gold, and the arcane claims, universal cures, and frequently harmful pharmaceutical remedies of the alchemists have made chemistry repugnant to all honest and sensible people, and produced a distaste for a knowledge characterized by fraud and deception" (Stahl, 1783: 2–3).

During this same period, a number of clearly-written books appeared that described chemical experiments in an accessible way. *Cours de chimie* (1675) by the French pharmacist Nicolas Leméry (1645–1715) was printed more than 30 times. In it, the author attempted to find some common ground between iatrochemistry and mechanical philosophy and his definition of a *principle* was well-regarded: "We are perfectly aware that these Principles are still divisible into an infinite number of parts that can rightfully be referred to as Principles. Therefore, we shall only use the term Principles of chemistry for those substances which have been separated and divided as much as our weak powers permit" (Leméry, 1682: 8).

The relationship between the corpuscular aspect of mechanical philosophy and the doctrine of elements remained problematic. How does one truly distinguish one substance from another? There was a need for something stable and enduring to link the invisible particles that were often described as either miniature hooks or joints (even pictorially represented as such by the Dutch physicist Nicolaus Hartsoeker in 1706) and the world that was accessible to the senses. Though the Stahl citation above represents a sharp break between Paracelsian nonsense and new "scientific" chemistry, it was Stahl himself who urged a *return* to a traditional science of Principles and Elements for fear of ending up in a blind alley if the program of the mechanical philosophers and the Newtonians, based on the absolute homogeneity of matter, prevailed. What was more, Stahl was a great admirer of the *Physica subterranea* by Joachim Becher (1635–82): he had a work of Becher's dating from 1669 reprinted under this title. He even referred to Becher as a great and irreplaceable master (Stahl, 1783: 5–7). What is so surprising about the *Physica* is that, in addition to a triple subdivision of the Earth which would prove important to both mineralogy and chemistry, it contains *every single* typical Paracelsian idea: the idea that the study of nature begins with a Mosaic account of creation; the analogy between macrocosm and microcosm; the parallelism between plants and animals; belief in spontaneous generation; the theory that metals "grow" in the bowels of the Earth; and even the parallelism between the perpetual and eternal circulation that takes place in the cosmos and chemical distillation.

In order to explain combustion, calcination, and respiration, Stahl again referred to Becher and introduced a principle of combustion called *phlogiston* to chemistry. The term *floghistòs* as an adjective means "inflammable," and dates to Sophocles and Aristotle (Partington, 1961–2: II, 667–8). Becher identified phlogiston or the inflammable element as the second earth element, and was more or less his version of Paracelsus' Sulfur or combustive element. The phlogiston theory was apparently satisfactory for explaining combustion and the calcination of metals (oxidation): a substance will burn if it contains phlogiston, and substances give off phlogiston during the processes of combustion and calcination.

As Ferdinando Abbri has shown, a phlogiston *theory* never existed. Until the conceptual breakthrough by Antoine Laurent Lavoisier (1734–94), the term phlogiston meant something different in every theory; it was a loaded concept, and functioned more or less as a "conceptual squeeze-box" (Abbri, 1978, 1984).

Phlogiston joins a long list that includes ideas such as the celestial spheres, the moving spirits of planets, *impetus* as a sort of internal motor, Cartesian vortices, caloric, female seed, animal magnetism, vital forces in physiology, luminous ether, nuclear electron. The history of science is rich with such ideas: once held to be true, confirmed by experience, and ferociously defended. These terms designated entities that have disappeared from the physical world and from today's science textbooks; they no longer interest scientists and are important only to historians of science.

11
Magnetic Philosophy

Strange Events

Sympathy and *antipathy*, "anthropomorphic" ideas that characterized the study of nature for millennia, seemed fairly obvious choices to apply to phenomena such as *attraction* or *repulsion*. The extraordinary and miraculous effects of magnets had been written about endlessly: stories of electric fishes that attach themselves to ships and slow their courses; magnetic islands that can wrench the nails from passing ships; and the protective power of magnets against witchcraft. Many of the popular beliefs about magnets were catalogued by Nicholas Cabeo, writing in 1629: how the smell of garlic can reduce or wipe out the power of a magnet; by interposing a diamond a magnet is prevented from attracting iron; the blood of a goat guards against any impediments to a magnet's powers; magnets can reconcile a husband and wife or reveal an adulterer; magnets can enhance one's powers of speech; they put one in favor with rulers (Cabeo, 1629: 338).

Lodestone is a variety of the iron mineral, magnetite, and has the unusual property of powerfully attracting iron. A steel needle that comes in contact with a piece of lodestone acquires the property of attracting particles of iron. If that same needle is free to rotate on a horizontal plane about its center of gravity, the same end always points north.

Amber, glass, ebonite, and sealing wax, when rubbed with a piece of silk or wool, attract bits of paper, hair, or straw. The term *triboelectricity* is used today for electrification resulting from friction, and we distinguish between *insulators* where a charge is limited to the areas of contact and *conductors* where the state extends to the entire surface area of the charged object. It was no easy task to introduce order and rules to the study of the magnetic phenomena which I have just described. This was a field in which strange things happened. For instance, an experiment that had been repeatedly successful suddenly and inexplicably failed when attempted on a humid summer day or in the presence of perspiring spectators. The first students of electricity had no awareness of the effects of either humidity or dryness. And the behavior of gems and precious stones, objects of great fascination to many early electricians, was just as capricious as that of glass. In fact, Newton even sent the Royal Society a message in December 1675 in which he emphasized the highly irregular and unpredictable nature of triboelectric phenomena (Heilbron, 1979: 3–5).

Mechanical models appeared inadequate for interpreting events which clearly seemed to involve *attraction, sympathy,* and *antipathy*. Trying to measure things that were not only ill-defined in size but also prone to persistent irregular behavior was no easy task. Mathematization, which had led to such successes in the fields of mechanics and astronomy, did not appear applicable to all things in the natural world. Kepler had studied Gilbert's book on magnetism, but like its author, resorted to qualitative analogy when stating that the sun possessed a motive and magnetic power, or a *soul*. Though Galileo supported Gilbert's results, he accused him of having vainly searched for the true causes of those results, offering his own "reasoning" as "conclusive proof." Galileo wrote that "[W]hat I might have wished for in Gilbert would have been a little more of the mathematician, and especially a thorough grounding in geometry" (Galilei, 1953: 406).

A good wish, but one made in vain. The methodological and theoretical gulf between mechanics and the study of magnetism, electricity, and heat was to last for some time still. Though some firm theories and methods of measurement had been established over the course of the eighteenth century, it was not until the end of that century that the science of electrology was established and *quantifiable concepts* (such as *charge, tension, capacity, potential, electric field*) were defined. Three of the greatest theorists in the field – the French engineer Charles Coulomb, the Englishman Lord Henry Cavendish, and the Italian physicist Alexander Volta – did their work in the final decades of the century, and died in 1806, 1810, and 1827, respectively. It is no wonder then that John Heilbron, the author of the finest history of electricity to date, devoted a bare 50 pages to the seventeenth century and over 300 to the eighteenth.

Gilbert

It is no simple task – and possibly pointless – to determine whether *De magnete magneticisque corporibus et de magno magnete Tellure physiologia nova*, written in 1600 by English physician William Gilbert (1540–1603), was the last example of Renaissance naturalism or the first of modern experimental science. *De magnete*, the first chapter of which reviews the most important works on natural magic, has been described as both. Gilbert's science was far removed from mathematical methods, and Galilean mechanics. He gave no measurements in the book, and its experiments were typically qualitative. In fact, his methods were basically like those of Giambattista della Porta, though the quality of his experiments was far superior: they were more ingenious, more detailed, and more carefully carried out. The scope of his work was also not much different from that of the treatise-writers of his day: to investigate "hidden causes" and the "secrets of things," to discover the "noble substance of the Great Magnet" and the medicinal properties of the lodestone. Gilbert preferred "true experiments and demonstrated arguments" to the "opinions and probable suppositions of philosophers." This was the basis for his experimental treatment of basic magnetic properties, which (except for the concepts of strength of a magnetic field and of the line of force, and the mathematical formulation) "does not differ essentially from the discussion of the subject in elementary modern textbooks of physics" (Dijksterhuis, 1961: 393). Because of his great mistrust of "professors," Gilbert turned to the work of an English seaman and professional compass-maker, Robert Norman (*fl.* ca. 1560–96). In 1581, Norman wrote a book about the declination of the magnetic needle, *The New Attractive, Containing a Short Discourse of the Magnet or Lodestone*. The book had been based on a lifetime of practical experience and was unknown within the scholarly community.

The encounter with technology was a significant one. Gilbert tried to use the measurement of the dip of a compass needle (with the aid of a complex map and quadrant) to establish latitude at sea. He believed this application to be a great discovery that would enable one to find latitude in cloudy weather "with minimal effort and a small instrument." Gilbert used globular lodestones which he called *terella* or *micro-earth* in his experiments. The first conclusion he reached was that the Earth itself is a magnet with magnetic poles that correspond to the North and South poles. Contrary to established belief, these geographical poles were not *geometric* points but *physical* ones. Just as the direction of a compass-needle was fixed, so was the direction of the earth's axis. Though Gilbert accepted the daily rotation of the earth because he believed that all spherical magnets were naturally capable of rotation, he did not accept the Copernican theory of the Earth's annual rotation around the sun.

Gilbert's second important conclusion was the clear distinction he made between electrical and magnetic attraction (he coined the successful term *Vis electrica*). He described magnetism (the lodestone's attraction to iron) as a sort of *coitio* or mutual coupling of two bodies and electricity (though he never used this term) as the *attraction* of small and light bodies to previously rubbed pieces of amber, jet, glass, resin and sulfur. The *versorium* he created was actually an electroscope.

Gilbert's accurate and ingenious experiments took place against a background of magic and vitalism. Matter was something both psychic and alive. Electrical attraction occurred through material effluvium while magnetic attraction (which is exerted even if another body is interposed) was a spiritual force, the action of a "unique and peculiar" *form* (not in the Aristotelian sense), a "primary, radical astral" form that resides in all globes – "the sun, moon, and stars" – and which, on Earth, is the true magnetic force that we call primary energy." In fact, the soul of a magnet was superior to that of a man. Earth, the *mater communis*, forms metals in her womb. The entire universe was alive and "all the globes, stars and even this glorious Earth have always been ruled by their own souls which have also been responsible for their self-preservation." Aristotle had made the mistake of giving a soul to the heavenly bodies but not one to Earth: "The state of the stars compared to the Earth would be dire if a soul had been denied to them and instead attributed to ants, beetles, and herbs" (Gilbert, 1958: 105, 309, 310).

Jesuits and Magic

Giambattista della Porta (1535–1615) devoted the entire seventh volume (of the 20–volume second edition) of *Magia naturalis* (1558 and 1559) to the wondrous uses of the magnet. And in the Italian edition published in 1611, he explicitly accused Gilbert of plundering his text and hiding his plagiarism behind insults and slander. Gilbert had indeed used della Porta's book (he cited him almost as much as Aristotle), but more as a framework than a source (Muraro, 1979: 145).

In *Philosophia Magnetica* (1629, Ferrara), the Jesuit Nicholas Cabeo (1596–1650) popularized the very themes that Gilbert had treated less than 30 years earlier. While rejecting the claim that the Earth was a magnet, he too tried to clearly differentiate between electrical and magnetic phenomena. He observed the presence of repulsion and attraction, and claimed that friction made it possible for fine effluvia to thin the surrounding air, and that the original density was re-established by the capacity of the effluvia to attract lighter objects. He remained skeptical about the extraordinary powers attributed to magnets. It was, however, a fellow Jesuit, Athanasius Kircher (1601–80), professor of mathematics, physics, and Oriental languages at the Collegio Romano, and an indefatigable polygraph, who claimed the title "Oedipus of his age." Kircher popularized the great ideas of his day. He established and organized a museum–laboratory of natural magic where he, on the one hand, fought to discredit the claims of alchemists and the builders of perpetual motion machines and, on the other, exhibited "magical machines" that produced optical illusions, communicated at great distances, or moved heavy objects without any visible means of transport; with the support of the Royal Society he examined if indeed tarantulas could be driven away by a ring of powder ground from the horn of a unicorn.

Kircher believed that Gilbert had been a great scholar of magnetism, whose only fault lay in the monstrous belief that the Earth did not stand still. He questioned Gilbert's claim that the Earth was a magnet, for if true, and a *terrella* the size of two hands attracts a pound of iron, then every horseshoe, piece of armour, pot, pan, knife, spoon and fork would cling so stubbornly to the ground that no force could possibly detach them. Man would not possibly be able to use iron. He also felt that Kepler had been a prince of an astronomer but wondered, if Kepler's world view in which the magnetic force of the sun could move planets were true, then why do compass needles not point toward the sun? (Kircher, 1654: 3–5, 383–6).

Book three of *Magnes sive de arte magnetica opus tripartitum* (published in Rome in 1641, in Cologne in 1643, and a second expanded edition in Rome in 1654) deals with terrestrial and celestial magnetism, the natural and artificial production of rain, the thermometer, the effect of solar and lunar magnetism on the tides, the magnetic force of plants, magnetism in medicine, the powers of attraction of music, love, and the creative mind (ibid.: 409).

The study of magnetism through experiments is but a specific example of an overall *vis tractiva* present in all things and distributed throughout nature. Therefore magnetism belongs not just to the magnet but to all natural things. Kircher repeatedly used an expression found in all books on magic: like is attracted to like and unlike is repelled by unlike. The connection between all bodily objects is the key to knowledge of the occult, more commonly referred to as *magic*, and which philosophers believe is the one, true knowledge (Nocenti, 1991: 180–9).

In the seventeenth century of the mechanical philosophers, Kircher represents the re-flowering of the curious and singular combination of magic and alchemy with modern experimental science. The magus and the mechanic appeared once again to be one and the same. Machines were seemingly built more for the purpose of exhibiting marvels and demonstrating wonders than to reinforce man's hold over nature. Kircher was not alone: in 1670, the Jesuit Francesco Lana Terzi, a student of Kircher's and a corresponding member of the Royal Society, wrote *Prodromo ovvero saggio di alcune invenzioni nuove premesso all'Arte Maestra* and another Jesuit, Kaspar Schott, wrote *Technica curiosa sive mirabilia artis libri XII* in 1664. Schott, who was read and admired by Leibniz, not only wrote about language and attractions but also about evil powers, multi-headed beasts, and demonic possession.

Hermetic Platonism was clearly being used in these types of texts as a sort of vindication. In this context, it would seem that Kircher's aim was to fulfill Francesco Patrizi's late sixteenth-century agenda of having the Pope replace the teachings of Aristotle with those of the Hermetic philosopher Marsilio Ficino. One wonders if the Jesuits had a specific cultural policy in mind when they produced work which mixed new with old superstitions and tended toward the sensational, the extraordinary, and the fantastic. Or was this output simply a typical manifestation of a Mannerist and Baroque mentality?

Cautious Experiments and Daring Devices

At the same time that Kircher was the center of great attention and a publishing success, Lorenzo Magalotti (1637–1712), secretary of the Accademia del Cimento, an indefatigable European traveler and Cosimo III's special ambassador to London, Sweden, and Denmark, published *Saggi di naturali esperienze* in 1667. His passion for precision and detached observation is greater than his curiosity in the strange and unusual. To go from Kircher to Magalotti is to enter another world in which the researcher is praised for restraint and caution, and "to experiment" is synonymous with challenges and obstacles, and knowledge is compared to a rough sea that is difficult to navigate: "An experienced experimenter knows first hand the difficulties one meets in conducting a laboratory experiment, if only in the use of materials [. . .]. Nevertheless, a multitude of wonderful magnetic actions have been discovered, and it is likely that there are many still to find, though we have not energetically plunged into the inquiry for we are all too aware that new discoveries require a great and lengthy study, uninterrupted by other investigations" (Magalotti, 1806: 163; 1976: 228).

Not all seventeenth-century discussions of electricity took place against a backdrop of magic and the occult. Magalotti was not alone in providing a contrast, as Cartesian mechanical philosophy and its penchant for model-building and systems overwhelmed and practically annihilated any focus on experimentation. For example, when Descartes' discussed magnetism in the *Principles* (1644), he never mentioned any of the detailed research (of the sort conducted by Gilbert) on individual magnetic phenomena. A strictly mechanistic perspective rejected any notion of *virtues* and *attraction* as magic or "occult." Magnetism had no effect on the motion of the Earth and other planets which were kept in motion by vortices of fine matter. All unexplained phenomena could be explained in terms of the Cartesian principles of size, shape, and motion. For example, iron filings clustered around the north and south poles of a magnet because particles of the first element, which had become channeled or grooved when squeezed between small spheres of the second element, moved through curved tubes or conduits. The "channeled particles," pictured as small snail-shells, pass easily through the Earth entering it at the north or south pole. Since the entire vortex rotates in one direction on its axis, those particles coming from the south pole must spiral in the opposite direction from those coming from the north. The grooved particles can easily pass through the earth because its interior is lined with grooved pores that accommodate the passage of right- or left-handed particles. Magnetic particles are able to penetrate the body of another magnet. Magnets come together because the particles are dragged through the intermediary air, and in the necessary absence of a vacuum, are forced to approach one another. They repel one another to make room for the flow of particles which, if like poles face one another, will not be able to enter the canals. Descartes was certain that "channeled particles could be invoked each and every time a case of attraction or repulsion occurred. This included electrical phenomena [. . .]" (Shea, 1991: 302–5). Cartesianism was alive and well in France until the 1740s (Heilbron, 1979: 31), and had such followers as Jacques Rohault (1620–72) and François Bayle (1622–1709).

The Sulfur Globe

Otto van Guericke, the author of *Experimenta nova* (1672), was a Copernican, and as such he was fascinated by the idea of a vast universe and void in which moved the celestial bodies. He felt certain that the artificial vacuum he had created in a famous and costly experiment (to be described in chapter 16) shared the same features of the interplanetary void. He also believed that the *virtues* or *powers* of the planets could be reproduced in a laboratory experiment. Guericke took a globe of glass the size of a child's head and filled it with powdered sulfur, heated it, let it cool and then broke the glass. He fixed the sulfur globe on an axis so it was free to rotate, and when rubbed it glowed and made crackling sounds, instantly revealing the very powers that belonged to Earth: it attracted light objects and made them hover while it rotated. The sulfur globe was the planet Earth itself under our very own eyes. It even exhibited a *vis repulsiva*, repelling attracted objects as a result of a conflict between different natures. The same thing happens when the Earth expels fire and other fiery substances and keeps the spherical body of the moon at a distance.

The only one of his discoveries that Guericke classified as *electric* involved the exercise of electrical power through a linen thread when one of its ends had come into contact with the electrified globe. The virtues (or effluvia) were at once corporeal and incorporeal. Incorporeal virtues included the impulsive, conservative, repulsive, directive or magnetic, and rotatory, in addition to heat sound, and light. The classification of virtues was complex and imprecise. Only his work with the sulfur globe was of interest to his contemporaries. Heilbron has pointed out that Guericke's description of electrical transmission along a thread remained obscure and had to be rediscovered before it could become a permanent part of the body of knowledge about electricity (Heilbron, 1979: 218).

Music and Tarantism

The scientific culture in which bizarre interests were pursued and experiments were conducted without the benefits of theories did not immediately respond to either Guericke's experiments or Huygen's meditations. Their effects were felt only when they were reconsidered in the middle of the next century, in a different theoretical context (Heilbron, 1979: 219, 226). As we have seen, times were very muddled indeed, but this did not mean that the dividing line between magic and science – so precisely drawn at the beginning of the century – had been forgotten. Descartes felt Kircher was more of a quack than a savant (Descartes, 1936–63: III, 803) and Evangelista Torricelli described Kircher's book to Galileo in this way: "This is a great big book about the magnet; a book attired in grand embellishments. In it there are astrolabes, clocks, and anemoscopes described in eccentric terms. There are many large and small beakers, epigrams, couplets, epitaphs, inscriptions, bits in Latin, Greek, Arabic, Hebrew and other languages. There is even a piece of music said to be an antidote for the venom of a tarantula. Enough: Signor Nardi, Magiotti, and I have had a very good laugh" (Galilei, 1890–1909: XVIII, 332).

Although the three friends did not have a satisfactory theory of magnetism and electricity between them, they had good reason to laugh. Unbelievable as it may seem, the thing that made them laugh the hardest – the musical antidote to tarantism – is the only item that 300 years later still arouses some interest. Ernesto De Martino's *La terra del rimorso* sheds some new and helpful light on the object of the men's laughter. He has studied the effect of music on victims of tarantula bites in southern Italy, and from this perspective discusses the significance of the imaginative work of Kircher.

De Martino astutely expresses the success and seductive effect of Kircher's ideas and of Hermeticism, still vigorous well into the seventeenth century. "In Kircher, that which bridges low ceremonial magic and Baconian knowledge as power was now being used to complete the inverse union with the astonished masses and to justify beliefs in traditional magic through the mental categories of natural magic. Kircher achieved a sort of counter-Reformation exorcism of natural magic; the attempt to provide a great synopsis of natural magic purified of any dangerous elements" (De Martino, 1961: 244).

12
The Heart and Generation

The Sun of the Microcosm

The study of physiology for anyone studying medicine in the sixteenth century (and much of the seventeenth as well) was based on the established and comprehensive view of the human organism that dated back to the Greek physician, Claudios Galen of Pergamon (ca. 120–200). Galen's system was not made obsolete by the work of the great Renaissance anatomists such as Andreus Vesalius, Realdo Columbus, Gabriel Falloppius, Bartolomeo Eustachio, and Fabricius of Acquapendente. According to Galen, the liver, heart, and brain formed a triad that both gave life and regulated it.

Since the arteries and left chamber of the heart in a dissected, blood-drained animal appeared empty, the Greeks had concluded that the arteries were full of air (as indicated by the Greek root of the word artery). Galen rejected this theory, though he did not believe that the blood circulated in a closed system. He defined *two circulatory systems*; the first system was made up of the veins and the right side of the heart and its purpose was to nourish the body. In this system, blood was produced by the liver which transformed food from the stomach and the intestines into venous blood. The second circulatory system was made up of the arteries and the left side of the heart, and its function was to transmit the "vital spirit" or "soul" found in the heart to all the parts of the body. Galen presumed that the intraventricular septum (the thick wall that divides the right ventricle from the left) was porous and allowed some arterial blood to pass into the left ventricle and mix with the air from the lungs, the function of which was to cool the heart and, through the process of respiration, rid the blood of impurities. The left ventricle receives air from the lungs; here the blood is enriched by vital spirits and transformed into arterial blood. According to Galenic physiology, the primary function of the heart was diastole, or dilation: in other words, the *attraction* of blood by the heart rather than its expulsion appeared to be the heart's most important function.

The great accuracy of the descriptions recorded by the great anatomists of the sixteenth century supplied science with a series of new facts. And these facts were truly innovative when integrated with the organic and coherent body of theory presented in *De motu cordis* (1628), a work by the English physician William Harvey (1578–1657). Harvey graduated in medicine from the University of Padua in 1602, and eventually (1651) was appointed professor of anatomy at London's Royal College of Physicians. He enjoyed the friendship of King Charles I who often attended his experiments. Harvey's house was ransacked during the Civil War and many of his notebooks destroyed. Never very interested in politics, he once confided to a friend that "the vacation from public business which causes tedium and disgust to so many has proved a sovereign remedy to me" (Pagel, 1967: 19fn).

The two greatest theorists of the mechanical philosophy – Descartes and Hobbes – praised Harvey's theory of the circulation of the blood as a major achievement. The doctrine became the springboard for the new mechanical biology and appeared to be the true subversion of Galenic physiology. Harvey was critical of Galen on a number of fundamental points: for example, the amount of blood pumped out of the heart in one hour exceeds the weight of a man and how is it possible for nourishment to account for the production of such a great quantity of blood? If the possibility of *continual circulation* is rejected, then where does all this blood come from and where does it end up? How can we prove that blood flows from the right ventricle to the left if the pores are invisible and therefore impossible to observe? Since the ventricular septum is harder and denser than many other bodily tissues, why was it (and not, for example, the spongy lung tissue) identified as a place where there is a flow of blood? Given that both chambers of the heart dilate and contract at the same time, how can the left chamber draw blood from the right chamber? And since animals lacking a pulmonary system also lack a right ventricle, is it not more reasonable to assume that the function of the right ventricle is to carry blood to the lungs? Finally, since we know that severing even a small artery causes the body to lose all of its blood in approximately half an hour, how can we assert that only some and not all blood circulates through the arteries?

Harvey took experimental data and questions and applied them to a new model in which blood circulated continuously and unceasingly through the body. According to this model, the primary process of the heart was not diastole but systole, or the moment the heart hardens and contracts to expel blood (like a pressing pump); the arteries pulse not because they themselves dilate but as a result of the pressure of the fluid that flows into them from the heart; the valves of the veins function to prevent a backflow of venous blood from the center to the extremities; the rich, hot blood that flows from the heart is depleted and cooled at the extremities of the body; blood travels from the ends of the arteries to the ends of the veins and constantly returns to the heart to give the body life. The arteries in the arm (and the limbs in general) are located deep under the skin's surface whereas the veins are more superficial. Harvey experimented by applying a tight tourniquet around the arm above the elbow and discovered that it prevented the arterial blood from flowing to the hand: the artery above the binding swelled, the hand got cold, and the pulse stopped. A moderately tight binding, on the other hand, prevented the blood from the veins from flowing back to the heart: the veins swelled under the binding, the hand became red and swollen, and pulse was weak but still perceptible.

It is important to put Harvey's *discovery* into context. He was obsessed with the problem of the purpose of circulation. As an Aristotelian he was predisposed toward the idea of natural circular motion. In Aristotelian philosophy, the circular motion of heavenly bodies insured the unity of the universe. This same principle guided Harvey in his study of the circulation of blood: namely the preservation of the body – the microcosm – through a continual circular regenerative movement of the blood. Moreover, the blood, insofar as it circulates through the body, was the bearer of the vital principle, or *anima* (Pagel, 1967: 25, 336). His insistence on the primacy of the heart, which he saw as "the sun of the microcosm," or a Supreme Ruler who governs the organism, is also reminiscent of Renaissance "solar literature," the greatest exponent of which was Marsilio Ficino.

Today we find the thought of an Aristotelian drawn to Hermetic ideas as disconcerting. Yet Harvey is more baffling still as he applied a mechanical model to classical knowledge and his own experimental data. Galen had compared the heart to a wick, the blood to the oil feeding it, and the lungs to the instrument fanning it, and supposed blood to be burnt and have a smoky waste (ibid.: 132). In this model, the arteries dilated as the result of a vital force and not because they were under pressure. Harvey's model was a hydraulic–mechanical one: the heart was like a pump, the veins and arteries like pipes that carry fluid, the blood a moving fluid under pressure, and the valves of the veins like mechanical valves.

This was the approach that allowed Harvey to reject the doctrine of spirits as it had been recast by the French physician Jean Fernel (1497–1559) in a popular physiological treatise, *Natural Parts of Medicine*

(1542). The arteries, left chamber of the heart, and brain cavities of a cadaver appear empty, and so Fernel concluded that in a living body they had been filled with an "ethereal spirit." The term *spirit* as it was used by Fernel and in Galenic medicine (which distinguishes between natural, vital, and animal spirits) struck Harvey as vague and unspecified, useless for empirical research, and altogether too mystical. Experientially, "we have never been able to find that spirit anywhere." Harvey was able to justify the notion of spirit by bringing it to a new level: the spirits were neither occult forces nor infinite powers that accounted for vital phenomena but rather the empirical features, characteristics, or qualities of the blood. Harvey had only had a glimpse of the process of the oxygenation of blood in the lungs and could only hypothesize the existence of capillaries to carry blood from the arteries to the veins. It was the English physician Richard Lower (1631–91) who later completed Harvey's theoretical work on oxygenation. And a microscope was necessary in order to actually *see* the capillaries, and so it was Marcello Malphighi (1628–94) in 1691 who first observed blood flowing through the capillaries into the lungs of a frog.

Malpighi, along with Robert Hooke, Jan Swammerdam (1637–80) and Antony van Leeuwenhoek, was one of the great *microscopists* of the seventeenth century. He became a member of the Royal Society in 1669. Between 1661 and 1679 he wrote a number of short works on the lungs, the tongue, the brain, the structure of internal organs, embryo formation in a chicken egg, and plant anatomy. These concise and clearly-written monographs embodied the search for the molecular structure of things, a search that utilized both the microscope and artificial processes such as cooking and dehydration (Adelmann, 1966).

We first encountered Alphonse Borelli in the chapter on mechanical philosophy. He attributed the motor abilities of muscles to a sort of chemical reaction between alkaline blood and acid nerve juices, and based his work on the theories of a Dane, Niels Steensen, who had used a microscope to study muscle fiber. Yet Borelli's attempt (like Descartes') to fully treat physiology in purely mechanical terms was only partially successful: beyond the mechanics of the skeleton and the muscular movement, the more complex questions of respiration and nourishment could not be answered by the fundamentals of seventeenth-century inorganic chemistry.

Ovists and Animalculists

The generation of living things was a topic of great debate in the seventeenth century (Roger, 1963; Solinas, 1967; Bernardi, 1980), and the contributions of William Harvey were significant. The motto *ex ovo omnia* appears on the title-page of *De generatione animalium* (1651), and he coined the equally famous saying, *omne vivum ex ovo*. In comparison to modern definitions, Harvey defined an egg as everything from the egg of a chicken or other oviparous animal to the cocoon and the amniotic sac of large mammals.

Francesco Redi (1626–98) provided experimental disproof of the ancient theory of spontaneous generation. According to this theory, insects and small animals such as flies, beetles, snails, leeches and some lower vertebrates can spring to life from dead and putrefied organic matter: cadavers produce worms, garbage begets insects, sour vinegar produces vinegar eels, and wasps and hornets are generated from the rotting flesh of horses, beetles from donkeys, and bees from cattle. Redi's method in *Esperienze intorno alla generazione degli insetti* (1668) was comparative, and he used what is known today as a control group. He put different types of meat in eight containers, sealed four of them and left four open. Larvae were observed only in the open four, which had been regularly visited by flies. It was immediately suggested that the absence of air was the reason for the absence of living forms. Redi then repeated the experiment and this time covered the four containers with gauze so that flies could not get at the meat and again no fly larvae were formed (Redi, 1668: 95).

Unexpected things happen in the history of science, like in any other history. What happened to Redi has justifiably been regarded as permanent scientific victory. Yet it was another equally important scientific discovery that appeared to threaten the very disproof of spontaneous generation. Antony van Leeuwenhoek (1623–73) lived all his life in Delft: he worked as an usher, could not read or write Latin, and was hardly capable of writing a scientific paper. He did, however, manufacture high-powered lenses and was insatiably curious about nature. Completely ignorant of what is known today as "scientific method," he looked at absolutely everything with his lenses. For more than fifty years he sent the Royal Society long letters written in Dutch, some of them accompanied by precisely detailed illustrations. He became so well-known that even Peter the Great visited him in Delft. In the summer of 1674, Leeuwenhoek discovered a host of tiny, swiftly moving, round, colored animals with tails in a sample of pond water taken near Delft. He found these small living organisms (protozoa) in many different kinds of water. Leeuwenhoek described this experience in a long letter to the Royal Society in 1676, and once the letter was published in the Society's journal, "Philosophical Transactions," what was one to make of Redi's claim that spontaneous generation did not exist? At best, Redi's claim seemed to apply only to that part of nature that could be seen with the naked eye.

Yet the power of the microscope showed that the limits to the existence of living organisms in nature were practically boundless. And Descartes himself had differentiated between the reproduction of higher-order animals (which he believed occurred by blending of male and female seminal fluids) and elementary life forms that are generated as a result of applying heat to matter. Those who supported the theory of spontaneous generation made use of Leewenhoek's discoveries to uphold traditional doctrines (Dobell, 1932).

Preformation Theory

The embryos of all mammals except monotremes (such as the echidna and the duck-billed platypus) develop inside their mothers and are nourished by the placenta. These animals are known as viviparous. Animals that deposit their eggs, such as birds, snakes, and fish, are known as oviparous. By the mid- to late-seventeenth century, the idea that viviparous animals also reproduced by means of invisible eggs was gaining currency. Redi had shown that even insects hatched from eggs. In *De mulierum organis generationi inservientibus* (1672), Reinier de Graaf (1641–73) confirmed Harvey's ovist theory. By the early 1770s, *ovism* was generally accepted theory even though mammalian "eggs" remained undetected until the early nineteenth century. In 1721 Antonio Vallisnieri had declared that the egg *must* certainly exist.

Leeuwenhoek announced his discovery of "spermatic animalcules" (spermatozoa) in a letter to the Royal Society in 1679. This time "little animals" were found in human semen. They had round, swiftly moving bodies with long, thin tails and a well-defined life-cycle. Why should these little animals be any different from those he had found in pond water? They originated in the testicles and he believed they were responsible for human reproduction. Leeuwenhoek also noted that there were more of these little animals present in the semen of one man than there were men on Earth.

Animalculism and its considerable following therefore challenged *ovism* and suggested that the origin of the preformed embryo of a mature individual resided in animalcules and not in eggs. Based on this theory, Leeuwenhoeck among others re-emphasized the difference between the reproduction of oviparous and viviparous animals. Animalculism was a difficult theory to accept: how could a tiny worm-like animal contain the human embryo? And why were the eggs of oviparous animals somewhat proportional to their body size while the size of an animalcule was more or less the same in different species? If every little animal potentially contained a perfect adult, how could the vast number of animalcules that never reached maturity be reconciled with the image of a divinely-governed cosmos?

In the early eighteenth century, ovist theory seemed to eclipse animalculism. However, both ovists and animalculists believed that an egg or a "worm" (male or female) contained a minute form (male or female) of its species. In order to better understand preformation theory (referred to in French as *emboîtement des germes*), it is important to realize that preformationism eliminated the need to consider the problem of how living organisms are formed, and transformed the question of *generation* into one of *growth*. An individual organism is not *potentially* but *actually* contained in an egg or sperm which is neither a blueprint nor a small-scale model, but a wholly-formed individual just waiting to be born. Fertilization simply activated the growth of an already fully-organized entity and caused its visible development to begin. The living thing was minute in size and hidden in either the sperm or the egg. Many scientists tried to find it with the help of the microscope, and Nicholas Hartsoeker (1656–1725) actually published a drawing of a minuscule dwarf with its head between its arms and its legs folded inside a "worm" (Bernardi, 1986).

Preformation theory eliminated the need to explain the origins of life in terms of vital forces or the ability of matter to organize, and was remarkably consistent with mechanical philosophy. However, the theory led to several potentially problematic conclusions. By allowing for only the growth mechanism in nature, and denying "forces" that organize the parts of an organism, then a chick which has been preformed inside an egg itself contains preformed chicks with their own preformed eggs. In *Recherche de la vérité* (1647), Nicholas Malebranche (1638–1715) gave a clear explanation of preformation theory. The seeds of living things have existed since the creation of the universe. They exist in miniature and are enclosed one within the other. The individual born one thousand years from now is just as perfectly formed as one born in nine months time. It is simply much, much smaller. In fact, Eve was in possession of all mankind, from creation until Judgment Day.

Preformationism was certainly a "strange" theory but, then again, the idea of infinite divisibility was compatible with ideas at the time about the infinite nature of things and theories of infinitesimal calculus. The theorists of calculus suggested that between one point and the next there exist an infinite number of points that form an infinitely divisible continuous line that is itself infinitely divisible into infinity: if such ideas were slowly gaining acceptance what would scientists in the late seventeenth-century find so shocking and unacceptable about a theory that is today considered strange?

13
Time and Nature

The Discovery of Time

Today we think of geology as the science of the origin, composition, structure, and history of the Earth and the organisms that inhabit it. We think of cosmology as the science that investigates the general laws of the universe and also deals with its origins and destiny. Geology and cosmology, sciences of the *vicissitudes* experienced by the Earth and the universe, are relative scientific newcomers and are linked to the profoundly revolutionary notion of the *discovery of time.*

Robert Hooke and his contemporaries in the 1630s believed that the world was six thousand years old; by the 1780s and 1790s, Kant and his contemporaries were aware that the Earth was several millions of years old. Was is it somehow different living in an age relatively close to the Creation (and with a Holy Book that chronologically outlined the *entire* history of the world) as opposed to one in which the past was – in the words of Count Buffon – a "dark abyss" of almost infinite time?

In the 100 years between Robert Hooke's *Discourse on Earthquakes* (1668) and Kant's *Universal Natural History and Theory of the Heavens* (1755), discussions about the history of the Earth and the history of the universe took radically divergent paths. The question was not simply one of different historical models, but could the histories themselves be subjects of scientific inquiry? If physics and natural philosophy dealt with the world *as it was* (as it was set in motion by God), then what sense was there in asking how the world had been "formed?" That question remained outside the scope of science and was relegated to the realm of groundless hypotheses or what we would today call science fiction. Only once it was considered legitimate to study nature from a "historical" perspective did two sharply divergent theoretical models emerge: history as a slow process of uniform and imperceptible changes (uniformitarianism) or history as episodes of violent catastrophes, measurable leaps and revolutions (catastrophism).

It was not always easy to distinguish between science and pseudoscience. Profound metaphysical assumptions were associated with the founding of cosmology and geology as scientific disciplines. Hooke, Descartes, Newton, and Leibniz did not just formulate theories; their inquiries had different ends and were oriented and limited in different ways. Those studying fossils and piecing together the history of the earth and the universe were constantly asking big questions: how did these things relate to the Biblical narrative and theology; what of the Creation and the Apocalypse; what position should be taken on materialism and Lucretianism; should one adopt an anthropomorphic or naturalist view of the world? Different ideas about science and different research methods significantly influenced not only the elaboration of theory but the "observation" of nature itself; the way in which we *view* some objects in the natural world (Rossi, 1979).

Strange Rocks

Were the abundantly available, odd little shell-shaped *rocks* unique stones produced naturally by some force of the Earth or the remains of primitive shells that had been transported to their present location by a flood, earthquake, or other such cause? Were the *lapides icthyomorphi* or fish-shaped rocks simply strangely-shaped rocks or were they the imprints of petrified fish? In the first case, the objects we call fossils were *viewed* as stones and natural objects that were "stranger" than other stones and natural objects found in nature. In the second case, they could be viewed as *documents* or *remnants* of the past, the evidence of changes and processes that had taken place at an earlier time. In the first case, the objects were simply observed and in the second, they were observed and "read" like documents.

To move from defining a *fossil* (from the Latin word *fodio* or "excavate") as any stone-like object found below the surface of the Earth to defining it as a relic or imprint of an organism that once lived on Earth required "discriminating the organic from the inorganic within a continuous spectrum of 'fossil objects'" (Rudwick, 1976: 44) as well as accepting the assumption that these curious objects could be explained in terms of their *origins* by interpreting them as vestiges or remains of the past. Once fossils were considered documents, nature (the realm of the immutable) could no longer be contrasted to history (the realm of action and change). Nature had a history of its own and shells were some of its documents.

With the exception of Leonardo da Vinci, who discussed the origins of marine fossils in both the *Atlantic* and *Leicester* codices, and Bernard Palissy (1510–90), the dominant interpretations until the seventeenth century were Aristotelian and Platonic. According to *De Mineralibus* (a spurious work), *fossilia* were formed as a result of a *succus lapidescens* or a bituminous wind that circulated inside the Earth's surface. The pseudo-Aristotelians believed that the sun's heat caused fumes to rise inside the Earth and this formed the metals and other fossils. The Platonic explanation of fossil formation appealed to forces or powers (*virtus plastica, lapidifica, vegetabilis*): an original "seed" gave birth to fossils which grew inside the Earth as living organisms. In *Meteorologica*, Aristotle described the Earth's interior as cracked, furrowed, and full of big, empty spaces. Earthquakes, he explained, were the result of solar-induced "winds" inside the Earth that made the Earth shake above the surface.

How are Natural Objects Made?

Robert Hooke (1635–1703) took a much broader view of *natural history* than his teacher, Francis Bacon. Up till then, the object of natural history had been to describe and classify natural objects, not to study the changes and alterations that nature had undergone through time. With respect to "shells," Hooke believed that science had to investigate "how, when, and under what circumstances bodies had been placed where they now lie." Not only was it difficult "to discern any sense of chronology from them," but the very chronology of nature was problematic. In *Discourse on Earthquakes* (written in 1668 but published after his death in 1705), Hooke returned to the question of fossils (first posed in 1665 in *Micrographia*) "that until now has deeply troubled all those devoted to natural history and philosophy." He deliberately distanced himself from both Aristotelian and Neoplatonic theories. He also rejected as improbable the theory that fossils dated to the time of the Great Flood. Hooke believed that the Earth and terrestrial life forms had a history. A series of *natural powers* and physical causes (earthquakes, floods, deluges, eruptions) had altered the Earth and its living organisms. From the time of the Creation, "a large part of the Earth's surface has been transformed and its nature changed [. . .] many parts that were not land now are, and other parts that are seas were once land, mountains became plains and plains became mountains." The Earth had originally been a fluid globe that slowly crystallized and solidified and is made up of superimposed layers. In order to explain the existence of fossils that belonged to no known species, Hooke had even renounced the idea of eternal and immutable species in favor of the hypothesis of the destruction and disappearance of living species: "Let us verify that changes in climate, environment, and food often produce great changes and there is no doubt that changes of this sort can produce enormous alterations in the shape and properties of animals" (Hooke, 1705: 334, 411, 290, 298, 327–8). Nevertheless, Hooke continued to place his "history" within the narrow confines of the Biblical narrative. He neither intended to reject the traditional 6,000 year chronology nor question the "agreement" between nature and Scripture.

In the middle of the seventeenth century the problem of *interpretatio naturae* was no longer based exclusively on spatial and structural dimensions but came to be linked to the dimension of time. To analyze and understand a substance involved more than just deconstructing it, reducing it to the movement of particles, and studying its geometric features. Other questions were becoming important: how did a natural object form over the course of time? how did nature produce, over time, a certain object? The Danish anatomist Niels Steensen (Nicholas Steno, 1638–86) outlined the terms of a new "theorem" of fossils with Cartesian precision. At the beginning of the *Prodromus* (1669), he states that "given a naturally-produced object of a certain form, one will find within the object itself the proof of how it was formed" (Steensen, 1669). Steno demonstrated strong Galilean and Cartesian tendencies in the *Prodromus*. The corpuscular theory of matter was used to make a clear distinction between a "crystal" and a "shell" or fossil. Though a Tuscan land study had been the basis for the hypothesis that the Earth's crust was covered by stratified layers that had formed as a result of the sedimentation of inorganic matter and fossilized marine remains, the idea came to be accepted as generally valid. The theory explained the presence of fossils found in the series of layers and represented a coherent attempt to reconstruct the sequence of geological events. The original horizontal position of the layers had been altered over the course of centuries by eruptions and earthquakes, and the ensuing cracking, sinking, and rising shaped the current landscape.

Agostino Scilla (1639–1700), painter and fellow of the Fucina Academy, published *La vana speculatione disingannata dal senso: Lettera responsiva circa i corpi marini che petrificati si truovano in vari luoghi terrestri* (1660) just a year after the *Prodromus*. Scilla, who had not seen Steno's work, countered the "vain speculation" that fossils had "grown" inside rocks with the proposal that their origins were organic. He claimed that fossils "had been real animals and not sports of nature previously generated simply from stone-like substances." He did not believe that minerals "grew" inside mines and ridiculed the "vegetability" theory of rocks. Each time we hold a fossil shark's tooth in our hands we are able to establish its exact location in the jaw of the shark (Scilla, 1670: 21, 26,

33, 86–7). Scilla constantly referred to the fact of his being a painter and advocated observation over speculation. He cited Lucretius and Descartes, and though he never mentioned Galileo by name, he relied on his theories. Aside from his sensism and skepticism, Scilla accepted only one philosophy: that which "recognizes the great disparity between what men think and what Nature has worked" (ibid.: 105). In 1696, William Wotton presented an abstract of Scilla's work to the Royal Society and the next year published *A Vindication of an Abstract of an Italian Book Concerning Marine Bodies*. In his own *Protogaea*, Leibniz compared the meticulous observations made by the "wise painter" from Messina to Kircher's fantastic visions of Moses and Christ on grotto walls and Apollo and the Muses in the veins of an agate.

Kircher's book, *Mundus subterraneus* (1664), had reached a large audience. His geological theory agreed with the Bible, and with regard to orogeny, he identified two different types of mountains: the one orthogonal to the Earth's surface and directly created by God and the other created by natural causes after the Great Flood. None of the fossils found in either type of mountain were the remains of organisms but were the fruit of a *vis lapidifica* and *spiritus plasticus*. In the Hermetic tradition, he compared the water circulating within the Earth to the bloodstream. The inside of a rock revealed traces of geometric figures, heavenly bodies, and letters of the alphabet that were symbols of divinity in nature. Kircher's tome intermingled themes from "chemical philosophy" and presented itself as an alternative to Descartes' mechanical explanations of the Earth and the universe.

The fossils that Colonna, Scilla, and Steno (whose book was translated into English in 1671) had written about were from the Holocene and Quaternary periods (subfossils, in modern geological terms). Since they did not present any significant differences from living organisms, it was easier, from the evidence available, to support the theory of the organic origins of fossils. The fossils in the collections of Martin Lister (1638–ca.1702), John Ray (1627–1705), and Edward Lhwyd (1660), dated to the Jurassic and Carboniferous periods, and in many cases were morphologically different from similar living species or (as in the case of ammonites) did not correspond to any extant species. Lister considered them rocks, and noting that fossils were not uniformly distributed but characteristic of particular strata, rejected Steno's geo-paleontological hypothesis. The forty days of the Great Flood apparently did not constitute a period long enough in which the strata of the Earth's crust could form. The theory of the organic origins of fossils highlighted the important differences being made between extant organisms and fossilized animals. The significance of these differences (for those who accepted that theory) necessarily led to the acknowledgment that some animal species must have died out. Extinction, however, appeared to contradict the "plenum" of reality and the great chain of being and introduced elements of imperfection and incompleteness in the work of the Creator. Certainly Lister, Ray, and Lhwyd rejected the theory of organic origins on the basis of technical difficulties and insufficient proof; however, there were also deeply-rooted metaphysical beliefs at stake.

A Sacred Theory of the Earth

In 1680 Thomas Burnet (1635–ca.1715) published *Telluris theoria sacra*, and a second expanded edition appeared in 1684. As he stated in the preface, his theory of the Earth could be called "sacred" inasmuch as it did not only consider the "common physiology" of Earth (as in the Cartesian view) but proposed to account for the *maiores vicissitudines* described in the Bible that form the "cornerstones" of divine Providence: the creation of the world out of a primordial chaos; the Great Flood; the final conflagration; and the complete destruction of all things. In order to preserve the universal character of the Flood and not reduce it (as the libertines would have it) to a chapter of local history, it was necessary to accept the Cartesian point-of-view that the Earth had once been different from today. In the beginning, there was a "fluid mass which contained all the materials and ingredients of all bodies haphazardly mixed together." The word of God transformed chaos into a world: the heavier parts plunged towards the center according to a descendent order of specific gravities, and the rest was divided up according to the same principle into one body of water and one of air. A process of sedimentation formed the first crust of the Earth which was perfectly smooth and devoid of rough spots and mountains. Inside the crust resided the waters of the "great abyss." This perfect surface, where the wind never blew and the temperature never varied, was the Garden of Eden. One world-wide cataclysm transformed this spherical Eden into the torn, troubled form we know today, with its oceans and jaggedly-edged continents. The sun's heat dried up the Earth, the crust split open, and a gigantic Earthquake cracked the surface of the world. The waters below then poured out to cause the Great Flood, and vapors from within condensed at the poles and rained down in torrents towards the equator. The axis of the Earth tipped with respect to the plane of the ecliptic, and this caused the changes in season and climate. When the floodwaters retreated into the great abyss (a slow process that was still occurring), the Earth was left in upheaval. The world was not evidence of God's original intention, but "a broken and confused heap of bodies placed in no order to one another." The moon and the Earth were "the image or picture of a Great Ruine, and have the aspect of a world lying in its rubbish" (Burnet, 1684: 249, 109). Burnet's depiction of rubble and the Great Ruin became something of a metaphysical *leitmotif*. The theme of ruin, associated with the idea of the slow corruption of the world and the progressive degradation of nature, was central to baroque and neo-Gothic culture. Burnet was attempting to reconcile Descartes' account of the creation of the world with the Bible. He believed that God had "synchronized" the events in the Bible with a chain of mechanical and natural causes. Burnet surely would have been surprised by the influence his book had on the history of the idea of the "sublime" and the rise of a reverie of Mountains. Burnet's ideas sparked a heated controversy. His book was constantly compared to Fontenelle's work on the plurality of worlds and was fiercely contested by the Newtonians. In *Geology or a Discourse Concerning the Earth Before the Deluge* (1690), William Temple (1628–99) obstinately countered every one of Burnet's claims with a passage from Scripture.

The image of a universe in progressive decline was incompatible with the Newtonian idea of the best possible world as the setting for God's work. In *Essay Towards a Natural History of the Earth* (1695), John Woodward (1665–1728), a fossil collector and physics professor at Gresham College who was expelled from the Royal Society for arrogance, rejected many of Burnet's claims and branded his history of the Earth as "imaginary and far-fetched." Many of the fossil remains found in England were of animals that lived elsewhere on the globe. According to Woodward, the Great Flood had caused the total destruction of the world, in accordance with Scripture: matter had been broken down into its basic parts, recombined, and separated anew. Fossils were the proof of this process. What emerged after the Flood was an environment congenial to human life, and so the changes and variations to the Earth's surface had been positive.

In *The Wisdom of God*, John Ray (1627–1705) insisted that nature was the manifestation of God's wisdom. The collection of waters in their great receptacles and the emergence of land were two such displays "because under these conditions water feeds and sustains many different types of fish in the same way that *terra firma* does for a great variety of plants and animals." William Whiston (1667–1752) was more ambiguous toward religious orthodoxy in the book he dedicated to Newton, *A New Theory of the Earth* (1696). Three cosmological theories are laid out in this work: (1) the Earth was formed following the cooling of a nebular comet with a mass equal to the Earth's but a volume far greater; (2) the Great Flood occurred when the Earth's interior waters burst forth following the Earth's convergence with the tail of a comet six times its size and 24 times closer to it than the moon; and (3) the Final Conflagration will occur either when the same comet or a new comet nears the

Earth, causing the waters to evaporate and the Earth to return to its original state. Edmund Halley (1656–1742), one of the greatest astronomers of his day, had already made a similar suggestion in 1694. His most important work, *Tabulae astronomiae*, was left unpublished in his lifetime (he feared being accused of atheism), and was published posthumously in 1742 in the *Philosophical Transactions*.

Leibniz and the *Protogaea*

The fate of Leibniz's *Protogaea* was a curious one. He wrote it between 1691 and 1692, more than a decade after Burnet's *Sacred Theory* and before the acclaimed books by Woodward and Whiston appeared in 1695 and 1696 respectively. However it was only published 56 years later in 1749, the same year that the first volume of Buffon's first edition of *Histoire naturelle* appeared. A two-page extract had been published in the Leipzig *Acta eruditorum* in January 1693, but aside from that Buffon was unacquainted with this work by Leibniz.

The *Protogaea* proceeds from several clear metaphysical assumptions that shape the history of the universe in three ways: (1) the history of the world is the unfolding of implicit possibilities present from the start and "pre-programmed" like in an embryo; (2) the "program" was selected by God, and it was not chaos that reigned in the beginning of the universe but divine Providence or the laws of general order so that the (best) of all possible worlds could arise; and (3) it is the limited perspective of man that views the history of the world as unfolding through change and unrest. Leibniz's grand view transformed all of the traditional terms of the question: mechanicism and finalism were not incompatible and it was possible to speak of the history of the world and the cosmos without being charged with the heresies of the libertines, atheists, and materialists. By relativizing chaos and disorder, Leibniz neutralized the positions of both the Cartesians and Burnet, and paved the way for an empirical investigation of past and present change in the history of the universe and the Earth.

Even the most upsetting and threatening of Burnet's theoretical outcomes was found tolerable. Though true that we "live atop ruins," they are not evidence of decay nor are they the documentation of a slow process of decline: "disorder occurred amidst order" and even the first, frightening disturbances created equilibrium. In the beginning, everything shaped by nature was regular, including the Earth itself. The folds and roughness came much later. If the Earth had originally been liquid, its surface would have been smooth and according to the general laws of bodies all solids are created by the hardening of liquids. The proof of this, as Steno had written, was solid bodies enclosed within a larger solid. such as, for example, "the remains of ancient objects, plants, and animals wrapped in stone." What now was a solid outer layer had been necessarily formed after the objects lying within it, "and moreover, it at one time had to have been liquid."

Leibniz began therefore by accepting the Cartesian framework and incorporating Steno's theories. Globes that had once been luminous and incandescent like the stars and the sun, became opaque bodies as a result of the waste produced by the incandescent matter. Heat had concentrated in the interior and the crust had cooled and hardened in a process resembling the workings of a furnace. If glass is produced by heating dirt and stones, then it is reasonable for "the great bones of the Earth, the naked rocks, the immortal silica to have been almost completely vitrified in that original fusion of bodies." Glass, which is the foundation of the Earth, is concealed in the particles of other bodies. These particles, corroded and split by water, underwent a number of distillations and sublimations until they produced a silt capable of nourishing plants and animals. During the cooling process, the hardening of the crust generated enormous bubbles filled with air and water. Depending on the material and temperature, the masses cooled at different rates and some collapsed to consequently form mountains and valleys. Rising waters from the abysses joined those flowing down from the mountains and caused flooding; this created deposits and the process was repeated, resulting in layers superimposed upon layers. Only the Earth's primitive or *base* rocks had been created by the cooling that followed the primordial fusion. Other types of rocks, as documented by the existence of strata, were formed by the re-concretion that followed precipitation-induced disintegration at different stages.

Leibniz understood that his discussion of the "dawn of the world" contained the seed of a "new science or natural geography." He believed he had found the general causes for "the skeleton, and so to speak, the visible bone structure of the Earth," formed by the Himalayan and Atlantic mountain ranges, the Alps and the great ocean beds. The structure presents elements of stability: it was the result of a process at the end of which "a more consistent state of things originating from the cessation of causes and their equilibrium is produced." Once this state was reached, all subsequent alterations were no longer the result of "general" but "specific" causes.

It has been pointed out that Leibniz was much less of a "diluvian" than many of his contemporaries (Solinas, 1973: 44–5). He emphasized Steno's theories both to explain the existence of layers (once horizontal, later inclined) and to account for fossil remains. He was convinced of the organic origin of fossils, and the arguments he made in favor of this theory and against his opponents were extraordinarily penetrating: "I myself have held fragments of rocks sculpted with a mullet, a perchis, an Argentine. I have witnessed the extraction of an enormous pike with its body bent and mouth open as if it had been buried alive and gone stiff under the gorgonian force of petrifaction [. . .]. Many take refuge, with regard to this matter, in the idea of *lusus naturae*, a meaningless term" (Leibniz, 1749: 29–30). Leibniz remained aloof from those who claimed that "the animals that currently roam the Earth were at one time aquatic and, once deprived of that element, slowly became amphibians until their progeny finally abandoned their ancestral homes" (ibid.: 10). This hypothesis not only contradicted Scripture but was highly problematic. Nonetheless, Leibniz did not exclude the possibility of changes occurring in animal species. Many people seem surprised by the presence of fossil remains of animal species "which we know it would be pointless to search the globe for [. . .] but it is plausible that animal species were greatly changed in the cataclysmic events of the Earth" (ibid.: 41).

Newtonians and Cartesians

Newton's theories of matter and the structure of the universe, presented by Richard Bentley (1662–1742) in the *Boyle Lectures* of 1691–92, were used as ammunition against Epicureans, freethinkers, and the champions of a popular millenarianism that followed in the wake of the revolution of 1668. Burnet's thinking bore links to that millenarianism while Newton's natural philosophy came to be largely used as an ideology. Bentley's lecture of November 7, 1692, entitled "A Confutation of Atheism from the Origin and Frame of the World," argued against "the atheistic hypothesis of the formation of the world" and equated the term *mechanica* with *fortuitous*. In the *Examination of Dr. Burnet's Theory of the Earth* (1698), John Keill (1671–1721), the first professor of Newtonian physics at Oxford and author of the celebrated *Introduction to true physics* (1700), harshly attacked "world makers," or the creators of imaginary worlds, and "flood makers," or the creators of imaginary floods. They presumed "to know the intimate essence of nature and to tell us exactly how God created the world, based only on the principles of matter and motion. Like pagan philosophers and poets, they were "crude, arrogant, and presumptuous." Their extraordinary cheek had been encouraged by Descartes, "foremost among the world-makers of our century."

Keill, like many other Newtonians, debased the great Cartesian cosmologies of the 1690s to the level of science fiction. He appealed to the importance of Newtonian science, the certainty of its laws, and the rigor of its definitions. Behind the Newtonians' opposition to the fantastic theories of the "world makers" and the appeal to Newtonian physics loomed three great assumptions that were not open to discussion: (1) the history of the Earth and the universe cannot be entirely explained by natural philosophy and included several miraculous events; (2) the veracity of the Biblical narrative cannot be doubted; and (3) it is necessary to acknowledge the presence, in nature, of final causes and entirely legitimate to adopt an anthropomorphic point-of-view, even in the realm of physics.

14
Classification

"Poa bulbosa"

Consider a small plant commonly found in European grasslands that has rough leaves and a greenish flower and that we describe today as belonging to the *Gramineae* family. Based on the system of classification developed by the great Swedish botanist Carolus Linnaeus (Carl von Linné) and still in use, we call it *Poa bulbosa*. This two-part name locates the plant within a *system*. Botanical (or zoological) *classification* or *taxonomy*, which has named more than one million animal and plant species (and a great number of insects and mites still await classification) is the discipline that handles classification; that unites different forms into progressively broader and more general groups: race, species, genus, family, order, class, phylum, kingdom.

If one is familiar with the structure of the system, the very name of the little plant contains an astounding wealth of information. The Linnaean system of *binomial nomenclature* is practical. Two words identify the organism: the first defines its genus and the second its species, distinguishing it from all others of the same genus. Just like, claimed Linnaeus, a first name and surname do for human beings. The identification of a species is not simply the identification of differences but also the recognition of similarities to others of the same genus. Latin names are used to avoid any problems with national languages. Linnaeus compared his system to an army divided up into legions, cohorts, platoons, and squads, and he conceived of it as a hierarchy of groups within progressively more inclusive groups. Each restricted level *progressively limits* the characteristics of organisms, while each more inclusive level includes an *ever greater number* of characteristics and like organisms. Every term corresponds to a place in the hierarchy. Imagine climbing up the inside of a funnel and finding oneself, at each new level, in the company of more and more creatures. The only company on the same level as our species (*Homo sapiens*) is the extinct *Homo erectus*. Ascending we find the genus *Homo*, then the family, *Hominidae* which includes the great apes and monkeys, then the order, *Primates,* characterized by an opposable thumb and large brain, then the class, *Mammal,* hot-blooded animals that suckle their young, then the *phylum cordata* that, at some stage have the same characteristics as vertebrates, and finally the kingdom, *Animalia*, which groups together all living organisms incapable of photosynthesis. Naturally, the same operation can be performed in the inverse order by *descending* the walls of the funnel.

At the end of the seventeenth century a great French botanist, Joseph Pitton de Tournefort (1656–1708), used seventy words plus a drawing to denote a geranium. For Linnaeus' *Poa bulbosa* he used 15 words: *Gramen Xerampelinum, miliacea, praetenui, ramosaque sparsa canicula, sive Xerampelinum congener, arvense, aestivum, gravem minutissimo semine*. We learn much more from Linnaeus' two-word description than from either Tournefort's seventy or fifteen word ones.

Classification

The popular perception of classification is that it is the slightly obtuse practice of giving Latin names to animals and plants. This is a caricature: "The quest of the best taxonomists has always been for a "natural" system, one that reflects the causes of order, not just an artificial arrangement for efficient pigeonholing" (Luria, Gould, Singer, 1981: 661). One of the most hotly-debated subjects in contemporary biology is that of cladistics or a system of classification that excludes any notion of "similarities" between living beings and operates solely on the basis of common evolutionary descent. It was the introduction of the question of evolution in the nineteenth century that complicated the work of classification. However, between the mid-sixteenth and early eighteenth centuries, the period covered by this book, classification dealt with a world of relatively fixed species, and fleas, flies, elephants, horses and giraffes were no different from the day they sprang from the hands of the Creator.

A number of problems need to be addressed separately: (1) the relationship, for the purposes of classification, of a theory of nature and a theory of language; (2) the fact that classification is not simply about knowledge but also memorization; and (3) the diagnostic function of classificatory language insofar as it must grasp the *essential* and ignore all the superfluous or incidental.

Universal Languages

In late seventeenth-century Europe, a number of efforts were made to create a "philosophical," "artificial," "perfect," or "universal" written language that would be capable (or so hoped the language theorists) of surmounting the confusion created by natural languages. It would be a language of "symbols" that referred not to sounds but directly to "things." For this reason, Bacon and Leibniz were especially interested in Chinese characters and Egyptian hieroglyphics. The pictorial representation of a thing should directly refer to the thing itself (in the manner of a computer icon or a road sign of two schoolchildren crossing a street). The image can then be understood independently of one's spoken language: it is *written* and *spoken* differently, but universally *understood* (regardless of language) in the same way. Why not use this as the basis for creating a written code and then an actual language? Wouldn't this finally free men from the confusion of tongues that was God's punishment of man for having built the Tower of Babel? (Rossi, 1983; Eco, 1995).

The work of George Dalgarno and John Wilkins, the leading theorists of a universal language, appeared in 1661 and 1668, respectively. The following is a list of some of their ideas that subsequently influenced seventeenth- and eighteenth-century plant and animal taxonomists:

1   There is a fundamental difference between a natural language and a philosophical or universal language. The system of symbols used in a universal language must be understood independently of the de facto spoken language, and its grammar must be different from that of a natural language.
2   The most important goal of a philosophical language is to create characters that draw on a commonly held mental image of a thing, and not the name for it that is currently in use.
3   The expressions used in a philosophical language must be "methodical": that is, they should be able to illustrate the connections and relationships *between things*.
4   The relationship between an expression and the thing it represents must be unambiguous, and to every thing and notion there must be assigned a distinct expression.
5   The project for a universal language implies a plan for a universal encyclopedia, or the complete and orderly enumeration of, in addition to the accurate classification of, all things and notions to which an expression or conventional "mark" must be assigned.
6   The creation of an encyclopedia is essential for the language to work, and requires the composition of *tabulae* (in the Baconian sense of the

word). As Descartes had already pointed out, since it is true that a perfect language requires the classification of all existing things, the limits of the encyclopedia are the same as the limits of the language.

7   The encyclopedia (even if partial) guarantees that every expression will also be a precise definition of a thing or notion. And a definition is precise when a sign indicates the exact *location* of a thing or notion within that ordered whole of natural objects which the tables of the encyclopedia reflect.

8   "The principal design aimed at in these tables," wrote Wilkins, "is to give a sufficient enumeration of all things and notions in order that the place of every thing may contribute to a description of the nature in it. Denoting both the General and the Particular head under which it is placed, and the common differences whereby it is distinguished from other things of the same kind. [. . .] by learning the names of things we will at the same time be educated about their nature" (Wilkins, 1668: 289).

A Language for the Discussion of Nature

The transition from the quest for a universal language to a project for the classification of natural objects was a smooth one, and it seems perfectly clear why one scholar claimed that Bishop Wilkins had proposed to "do with words precisely what Linnaeus was later to do for plants" (Emery, 1948: 176; cf. Rossi, 1984). Yet this was no coincidence, not a simple analogy between the thinking of a "linguist" and of a botanist. Wilkin had turned to both John Ray (1627–1705), one of the great founders of botany, and the zoologist Francis Willughby (1635–72) for help in 1666 so he could include in his book "an accurate enumeration of all plant and animal families" (Ray, 1718: 366). The subsequent discussion between Wilkins and Ray is interesting on a number of points.

In 1682, the Reverend John Ray published the *Methodus plantarum nova*, followed in 1686 by the first folio of a massive work, the *Historia Plantarum* (1686–1704): 3,000 pages divided into three substantial folios describing 18,000 species and varieties subdivided into 33 classes based on morphological features. Ray was the first to differentiate between single and double lobed plants and introduced the modern definition of a species as a set of morphologically similar individuals derived from an identical seed. Ray, however, was a man of many interests. In addition to botany, he also wrote on theology, the Great Flood and fossils, on rhetoric and science, and the much-debated question of the superiority of the moderns. In 1674 and 1675 he published two dictionaries: *A Collection of English Words not Generally Used* and *Dictionariolum trilingue*. His interest in language was not that of a dilettante, nor was his interest in Wilkins's project superficial. He in fact undertook the thankless task of translating all of Wilkins's *Essay* into Latin so it would be accessible to scholars throughout Europe; he actually completed the project, though it was never published (Ray, 1740: 23).

Wilkins's much-hoped for "perfect" classification struck Ray as impossible to achieve. Ray did not believe, as Wilkins had, that nature could be geometrically and symmetrically arranged. For instance he did not believe that each of three classes of grasses could be further subdivided into nine "differences." In the face of such a request, Ray observed that nature made no leaps; it produced intermediate species that were difficult to classify, and its seeming continuity was the outcome of a series of imperceptible steps.

To Name is to Know

"To accurately execute this plan," wrote Wilkins, "it is vital that the theory that underlies the plan precisely draw on the nature of things" (Wilkins, 1668: 21). If one understands the characters and *names* of things, one will also learn about the *nature* of things. If the relationships, comparisons, and connections between terms in a terminology reproduce the relationships, comparisons, and connections between the things they name, then naming is the equivalent of knowing. As Linnaeus succinctly put it, *Fundamentum botanices duplex est: dispositio et denominatio* (Botany has a double foundation: to organize and to name) (Linnaeus, 1784: 151).

Joseph Pitton de Tournefort (1656–1708), a botany professor at the Paris Jardin des Plantes, based his system of classification on genus alone. Almost 700 types and more than 10,000 species in his works are described in *Eléments de botanique ou méthode pour reconnaître les plantes* (1694) and *Institutiones rei herbarie* (1700). Tournefort was equally confident that the characteristics or distinguishing properties of a plant were so intimately bound to its name as to be inseparable from it. He believed that botany was much more than the study of the properties of plants, and should not be considered simply the handmaiden of pharmacology and medicine. Galen and Dioscorides had made great contributions to medicine but had neglected botany by randomly naming plants one by one as they were being discovered. To elevate botany to the level of a science it was instead necessary "to begin the study of plants by means of the study of their names" (Tournefort, 1797: I, 47). He knew that a truly precise and perfect language would require the radical overturning of existing terminology and was mindful of tradition and a store of historical terms: "If plants had not yet been named, it would be useful to learn about them through simple names whose endings would indicate the relationships between plants of the same genus and class [. . .]. Doing this would require turning all of botanical terminology upside down; in the beginnings of this science such accuracy was not possible given that one had to name a plant simultaneously with its discovery" (ibid.: I, 48). So botany was far from perfect because of a sort of original sin: "For whatever evil fate, the more the ancients showed the many uses of medicine the more they neglected botany. In fact, they invented new descriptive names for plants according to their virtues without yet having rules by which to name them in anything but an arbitrary way" (Tournefort, 1700: I, 12–15).

Mnemonic Aids

In Bernard de Fontenelle's eulogy of Tournefort at the French Académie, he observed that Tournefort had given "order to an infinite number of plants scattered haphazardly upon the Earth and under the seas, and made it possible to categorize them by type and species in such a way that they could be easily remembered, and so the memory of botanists would not collapse under the weight of an infinite number of names" (Fontenelle, 1708: 147). Many believed that classification was an invaluable mnemonic aid, the necessity of which – as an integral part of the new scientific method – had long been championed by Francis Bacon. For in spite of his criticisms, Bacon had drawn widely on the rich tradition of the Ciceronian art of memory (cf. Rossi, 1983; Yates, 1966).

Fontenelle was not the only one to suggest that memory could be taxed by the weight of information. As we know, between the middle of the sixteenth and middle of the seventeenth centuries there was a dramatic change in the general condition of the natural sciences, and this encompassed *quantity* of information as well. The *Herbarum vivae icones* (1530), written by the great German botanist Otto Brunfels (1439–1534) and masterfully illustrated by Dürer's pupil, Hans Weiditz, recorded 258 different plant species. Just 100 years later, the Swiss naturalist Gaspar Bahuin included almost 6,000 species in the *Pinax theatri botanici* (1623), and John Ray, as we have seen, dealt with 18,000 species in his work.

The situation in the natural sciences was indeed complex, and confusion ruled the day. In his German translation of Linnaeus, Johan Friedrich Gmelin presented 27 different systems for the classification of minerals that were in use throughout Europe from the middle of the seventeenth to the middle of the eighteenth century. With regard to the history of their science, the followers of the Swedish master all agreed on one point; that Linnaeus had most importantly put an end to an age of *confusion.* One Russian disciple wrote that "the science of nature had been little cultivated before the last one hundred years [. . .]. With regard to the classical period, I am afraid that the descriptions of natural objects I have come across here and there are so lacking as to be rather useless. Memory alone is not sufficient for so great a number of objects, and classical authors had neither established a definite terminology nor arranged the objects in any order or according to any system" (Linnaeus, 1766: VII, 439).

The Essential and the Incidental

To better understand how the type of classification associated with universal language and Aristotelian classification differed, let us take a look at the ambitious attempt by Andrea Cesalpino (1519–63) at the end of sixteenth century to establish a botanical–zoological science on the Aristotelian principles of matter and form. In *Sixteen Books on Plants* (1583), he argued that plants, like animals, contain a living principle, and are simplified versions of more complex organisms. In other words, a plant is an upside-down animal with its head in the ground: its roots are the mouth through which it feeds; the fruit is the equivalent of an embryo; the sap is like blood. Gesner (1516–65) should also be remembered in this context: the *Historia animalium* listed animals alphabetically by their Latin name in 4,500 folio pages. Among the many splendid engravings illustrating the text was Dürer's famous rhinoceros head.

Quite a few historians and many epistemologists have completely ignored the meaning, the immensity, and the importance of the huge accomplishment of *tabulating* natural objects by botanists, zoologist, mineralogists, and in general, all scholars of "natural things" in the seventeenth century.

To capture the essential and ignore the superfluous: where does one look for the essential, and how does one distinguish it from the superfluous? Classical and Renaissance treatise writers discussed at length allegories, myths, and legends associated with a particular plant or animal, the subject's suitability as food, possible uses for it, and its poetic and literary representations. In the seventeenth and eighteenth centuries, botanical and zoological works relegated the so-called *literary* part to the end of a book where it became something of a curious appendix. Though the English physician and naturalist John Johnston (1603–75) still placed the unicorn together with the elephant in *De quadrupedis* (1652), he cut most of the literary discussion that had been included by Ulysses Aldrovandi (1522–1605) in his voluminous natural history.

The quest for the essential naturally took a number of different paths.

The 17 classes of plants in John Parkinson's *Theatrum botanicum* (1640) include the fragrant, poisonous, narcotic and harmful, soothing, and warming, as well as umbellifers, grains, swamp plants, aquatic and marine plants, trees and fruit-bearing trees, exotic and uncommon plants. Tournefort divided trees, shrubs and grasses from one another (a distinction which Linnaeus rejected) and subdivided them on the basis of the corolla, though he also made use of the differences between their fruits, leaves, and roots. Differences based on medicinal uses and place became outmoded. The approach taken by Linnaeus – classification based on the reproductive organs of plants – was a difficult one because plant sexuality had been rejected by such scientists as Malpighi and Tournefort, and was not generally accepted until the first half of the nineteenth century. Even more complicated was the situation for the animal kingdom. Linnaeus classified Mammals, Birds, Amphibians, and Fish as red-blooded animals, and Insects and Worms as the two classes of white-blooded animals. While true that Linnaeus was the first to classify man as an animal, it is equally true that he placed him among quadrupeds together with anthropomorphic monkeys and the three-toed sloth. Moreover he believed the rhinoceros was a rodent, and that the amphibian class included the crocodile, turtle, frog, and snake as well as the sturgeon and skate. He placed squid, cuttlefish, and octopi in the Worm class.

15

Instruments and Theories

Sensory Aids

To see, in the modern scientific sense of the word, means almost exclusively *to interpret signs generated by instruments*. More than a dozen complex devices come between the astronomer using the Hubble telescope and those distant galaxies that excite the astrophysicist and spark the imagination of all: a satellite, a system of mirrors, a telescopic lens, photographic equipment, scanners that digitize images, computers that control the photography, scanning and storing of digital images, a device in space that transmits the images to earth as radio signals, a device on the ground that translates the radio signals into computer code, software that reconstructs and colorizes images, video equipment, color printers, and so on (Pickering, 1992; Gallino, 1995).

In his book on the philosophy of science, *Representing and Intervening*, Hacking argues that in order to understand what science *is* and what science *does*, the two terms must be taken together. Science is made up of two basic activities: theory and experiment. Theories try to say how the world is; experiments verify theories, and the subsequent technology can change the world. We represent and we intervene. We represent in order to intervene and we intervene in light of the representations we have created. From the time of the Scientific Revolution a sort of *collective methodology* has given free rein to three basic human interests: speculation, calculation, and experimentation. The collaboration between the three has enriched each area to an extent otherwise impossible (Hacking, 1983: 31, 246). Francis Bacon taught that science is not just the observation of nature in the raw. Man's senses are heightened through the use of instruments. Newton's rays of light, like the particles of contemporary physics, are not facts of nature but instrument-induced "facts." Bacon urged that in the face of nature we must also "twist the lion's tail." From this perspective, the history of scientific instruments is not external to science but a basic and integral part of it.

Such an example can be found in the seventeenth-century debates on the weight of air and the existence and nature of the void. In the fourth book of the *Physics*, Aristotle defined space as the immobile limit surrounding a body and rejected the void. He argued that motion in a void is impossible for, according to his mechanics, if it were it would have to be either instantaneous or infinite. Moreover, bodies would fall at the same speed regardless of their weight. Expressions such as "nature abhors a vacuum" and "*horror vacui*" began to appear in thirteenth-century texts and became commonplace terms. Just as the Cartesians (who defined matter as extension) would later assert, when matter in a cosmic plenum moves from one place to another it is immediately replaced by other adjoining matter. Seventeenth-century philosophers were instead introduced to a different ancient view of the void, the opposite of Aristotle's, in the work of Diogenes, Cicero, and in particular, the *De rerum natura* by Lucretius. Lucretius (whose ideas had a great influence on Giordano Bruno) championed the view that a vast number of worlds were scattered randomly throughout infinite space. Even the Stoics, in the words of Simplicius, believed that one spherical, full and finite world was surrounded by a three-dimensional void empty of worlds and matter (Grant, 1981).

In Florence in 1644 Evangelista Torricelli instructed his student Vincenzo Viviani to fill a tube with mercury, close it at one end, and invert the open end of the tube in a dish of mercury. The level of the mercury in the tube sank until it was about 760 millimeters above the level of the dish leaving an empty space at the top of the inverted tube. What determined the level of the mercury to change? Did air have weight? And what was the nature of the "empty" space in the tube above the level of the mercury? This experiment with mercury is still known as "Toricelli's barometric experiment."

A number of different answers were proposed in the years between 1645 and 1660. The Peripatetics rejected the possibility that air had weight and that void space was possible; they suggested that a tiny amount of air remained in the tube and this expanded to its fullest when the level of mercury in the tube dropped. Descartes and his followers accepted the notion of air pressure but rejected the void; they believed that the space above the mercury was full of a matter so fine that it could penetrate the glass of the tube. The staunch anti-Cartesian, Gilles Personne de Roberval, accepted the void but rejected the possibility of atmospheric pressure, or the weight of air.

Torricelli had observed that the height of the column of mercury was subject to variation and so suggested that the device could be useful in measuring atmospheric pressure. At more or less the same time in Rouen, France, young Blaise Pascal – who had already published an essay on conics and invented the first calculating machine – began experimenting with glass tubes of such different lengths and shapes that they could only be supplied by the local industrial glassworks. His experiments proved that the height of a column of mercury remained invariable, and by doing this he disproved the hypothesis that the volume of empty space remained constant because the air that was left in the tube had reached its maximum degree of rarefaction. He fully exploited the technical capacity of the glassworks to produce large pieces; in one experiment he took tubes 14 meters long and secured them to movable ship's masts, filled them with water or red wine and inverted them in containers filled with the same. He also designed an experiment still written about in contemporary textbooks on physics: what happens to a column of mercury that is measured at the base of a mountain, during its ascent, and finally at the peak? The experiment, called *La grande expérience sur l'équilibre des liqueurs*, was executed on September 19, 1648 by Pascal's brother-in-law Florin Périer on the Puy-de-Dôme in Auvergne. A month later, Pascal published a detailed account. In 1647, Pascal published *Expériences nouvelles touchant le vide*, and in 1653, *Traité sur l'équilibre des liqueurs*. In 1647 he still believed that nature abhors a vacuum but by 1648 he suggested that gravity and air pressure – not a horror of the void – were the true causes of his experimental results.

The experimental research of Otto von Guericke and Robert Boyle also contributed significantly to settling the questions about voids and atmospheric pressure. In 1654 von Guericke, the mayor of Magdeburg, carried out a spectacular experiment before the Diet assembled at Ratisbon. Two brass hemispheres about 24 centimeters in diameter were brought together to form a sphere and the air inside it was evacuated. The hemispheres were then separated, but only with difficulty, by two teams of four horses. It took 24 horses to separate two larger hemispheres in a subsequent experiment. Robert Boyle designed his own experiment of "the void within the void" (which Pascal had also undertaken), in which he took an apparatus similar to Torricelli's, marked the level of the mercury, and enclosed the device in a container from which the air was gradually pumped out. Because of the drop in air pressure in the container, the level of mercury slowly dropped. Boyle did not call the vacuum he created in his experiments a *void*. He had no desire to be labeled a "vacuist" or a "plenist". Was the empty container devoid of any "subtle or ethereal" matter? Boyle was hesitant to pronounce on this, and believed it was more a question of metaphysics than physics and therefore outside the realm of "experimental philosophy" (Dijksterhuis, 1961: 457;

Shapin and Shaffer, 1985). As Dijksterhuis describes it, "it is true that they had freed nature of the horror vacui, but the latter, [. . .] had subsequently seized upon their own minds. The numerous ether-theories, which were to occupy such a prominent place in physics, are striking evidence of it" (Dijksterhuis, 1961: 457).

Torricelli's barometer was one of the six great scientific instruments of the seventeenth century (the microscope, telescope, thermometer, barometer, pneumatic pump, and precision watch) which were intimately bound up with the advancement of learning.

Intellectual Aids

As noted earlier in the chapter, science in the wake of the Scientific Revolution was characterized by theory, calculation, and experimentation. And the most powerful theoretical tool in the history of mankind was undoubtedly the new mathematics that emerged between the middle of the seventeenth and the early eighteenth century.

At the heart of the new mathematical method were the questions of infinity and continuity. In the process of calculating the distance between Mars and the sun at different points in its orbit, Kepler realized that his principal mistake was believing that a planet's orbit was a perfect circle. Indeed he declared that a *pernicious error* of great authority had made him waste valuable time. To what was Kepler alluding? According to the ancients, the perfection of the circle rested on the fact that every point of the circumference is both an end and a beginning. Similarly, there is no perceptible difference between beginning and end in a straight line and straight-line motion is never completed. Aristotle distinguished between potentiality and actuality and only recognized *potential* infinity. Infinity is not real, either as a reality in itself or as an attribute of something real. Infinity cannot be attributed to any thing or its parts. When we say that time or numbers are infinite or that a continuum is divisible to infinity, we simply mean that, for example, the action of dividing or adding numbers to numbers can be repeated an infinite number of times. Aristotle believed that the path to infinity consisted only in the infiniteness of the path (Wieland, 1970).

Let me take a moment to clarify why the modern concept of infinity differs so from the notion of an endless addition of one thing to another. Infinity and continuity were conceived of differently after the Scientific Revolution. The succession of one number to the next is *discontinuous*. On the other hand, the infinite succession of points in a straight line is *continuous*, and it makes no sense to speak of one point following *immediately* after another; instead an infinite number of points forming a continuous line segment which is itself infinitely divisible into continuous segments exists between one point and the next. In moving from one point to the next, one moves through an infinite number of points; a true infinity rather than one that might be constructed; an actual rather than a potential infinity (Lombardo Radice, 1981: 12). A continuous line was conceived of as divisible into an infinite sum of indivisible parts.

It may be that the Greeks were frightened by the idea of a process without end; they certainly found what we call the paradoxes of the infinite to be insurmountable barriers (Kline, 1953: 57). The method of exhaustion used in geometry to calculate areas avoided confronting infinity and infinitesimals as the object of a proof: to prove that a given figure has a given area one supposes an area either greater or smaller and then considers a series of inscribed or circumscribed figures that are closer and closer to the area of the given figure until one reaches the conclusion, *ad absurdum*, that the area can be neither greater nor smaller. The same process was used to calculate the area of both solids and volumes.

In 1615 Kepler cleverly answered the following question: why is it that the shape of a wine barrel, relative to capacity, is the most economical in terms of amount of lumber used? He adopted infinitesimals rather than the method of exhaustion to solve the question of the volume of a wine cask and why that particular shape used the least amount of lumber possible. He broke the figure into infinitesimal parts but did not discuss the significance of his method. Galileo Galilei, whose work contains atomistic ideas, claimed that as a continuum can be divided into parts that are themselves necessarily divisible, then it follows that it must be made up of an infinite number of dimensionless indivisibles. This was not a purely mathematical problem: Galileo was connecting the existence of dimensionless indivisibles with the composition of liquids and the phenomena of condensation and rarefaction. In the *Discourse* he asked if it was possible to compare two infinite sets by calling attention to the paradox that results when comparing the infinite set of whole numbers and the infinite set of their squares. Write the numbers 1, 2, 3, 4, 5, etc. on one line and below it the numbers corresponding to their squares, 1, 4, 9, 16, 25, etc. There is clearly a one-to-one correspondence between each whole number and its square. It is just as clear that the first series should be larger than the second for it contains both squared numbers and non-squared numbers. How then can one infinite quantity be larger or smaller than another? Galileo was forced to face one of the fundamental properties of infinite sums: a part of a sum can be the same size as the entire sum. In the *Discourse*, Salviati correctly concludes that the set of perfect squares is not smaller than the set of whole numbers but he did not go so far as to conclude that the two were equal in size. Galileo decided that man's finite intelligence could not discourse upon infinites, "the attributes of equality or greater and lesser than have no place among infinite quantities but belong only to defined quantities and [. . .] they are not suited to infinite numbers for one cannot say that one is greater than, less than, or equal to another."

Two of Galileo's students, Bonaventura Cavalieri (1598–1647) and Evangelista Torricelli (1608–47), resolved to steer clear not only of atomistic philosophies but also too-weighty philosophical positions. They did, however, manage to face "the vast ocean of indivisibles" in geometry precisely by comparing infinites. An area, for example, was made up of an infinite number of parallel segments, and volume was an infinite number of parallel planes: segments and areas themselves were the indivisibles. One could now find area and volume by comparing, one by one, the *indivisibles* that made up those areas and volumes. The men never actually stated that an area is the sum of infinite lines or that solids are the sum of infinite surface areas, but limited themselves to stating that a surface area is to the union of line segments as a solid is to the union of its sections.

The ancients used geometry to solve all arithmetic and algebraic problems. The use of negative roots to solve cubic equations, as Niccolo Tartaglia (1506–57) and Gerolamo Cardano (1501–1571) had done, would have been unacceptable to Greek mathematicians insofar as they were quantities that could not be expressed geometrically. In the section on mathematics in the chapter on Descartes, we learned that the "translation" of geometrical concepts into algebraic ones was an important development in light of the mathematization of physics. In analytical geometry, problems are solved by means of algebraic equations and equations represent curved lines. Every curved line that belonged to the science of mechanics and had been overlooked by classical geometry (insofar as they could not be drawn with a ruler and compass) was now at the center of scientific research. Between the seventeenth and nineteenth centuries, algebra occupied a distinctly superior position to geometry.

Infinitesimal calculus, or in Newton's words the method for calculating fluxions, was capable of finding the area inside a curve, solving the problem of tangents, and dealing with problems of continuous motion. Pierre Fermat (1601–65) in France, Isaac Newton (1642–1727) in England, and Gottfried Wilhem Leibniz (1646–1716) in Germany were all working on the same types of problems. In fact, it was the discovery of the calculus that led to one of the greatest scientific disputes in the history of science. The quarrel between Leibniz and Newton over priority of discovery divided scholars then and remains a subject of study today (Hall, 1980; Giusti, 1984, 1989).

In the *Principia*, Newton used classical geometry to prove his theorems (even the ones solved with calculus). He did not have the mentality of a pure mathematician for he viewed mathematics in terms of physics and believed that the very calculus he had invented was instrumental and "practical" in nature. He had, however, carefully studied the second edition (in Latin) of Descartes' *Geometry*, the algebra of François Viète (1540–1603), and the mathematical writings of John Wallis (1616–1703). Perhaps it was precisely because he lacked a solid background in classical geometry that Newton clearly saw the importance and essence of analytical geometry: curves and equations correspond to one another and the equation expresses the nature of a curve (Westfall: 1980, 107). In *De quadratura curvarum* (1676) he rejected Cavalieri's method and treated mathematical quantities not as if they were made up of arbitrarily small quantities but generated by continuous motion. Lines were described not by the adding up of parts but as the continuous motion of points; surface areas as the motion of lines; solids as the motion of surface areas; angles as the rotation of sides; and time as continuous flux. These operations "truly occur in nature and can be observed daily in the movement of bodies." In equal time periods, the quantities generated by those movements depend on the greater or lesser of these movements. "Considering that the quantities generated," wrote Newton in *Tractatus de quadratura curvarum*, "in the same time are larger or smaller according the greater or lesser velocity at which they grow, I sought a method to determine the velocities of motions and the increments by which they are generated; calling the velocities of growth *fluxions* and the quantities generated *fluents* or flowing quantities; in 1665–6 I gradually arrived at the method of fluxions which I use here on the quadrature of curves" (in Castelnuovo, 1962: 127–8). Fluxions "can be considered with arbitrarily great approximation as the increments of fluents generated during arbitrarily small intervals of time." Knowledge of the quantity allows determination of velocity; and knowledge of the velocity determination of the quantity. The speed of growth (or rate of change) is nothing but the (derived) fluxion of a (variable) fluent. The search for the relationship between fluxions and fluents (today known as integration) seemed much more difficult to Newton than finding the relationship between fluents and fluxions (known as derivation) even though he was perfectly aware that derivation was the *inverse process* of integration (Singh, 1959: 34; Giorello, 1985: 172–3).

We now have some key concepts beginning with *instantaneous velocity*. This term was *not* defined as velocity divided by time. The "calculus" introduced instead the concept of the number approached by average speeds as the intervals of time over which the average speeds are computed approach zero (Kline, 1953: 220). Leibniz and Newton independently arrived at the idea of taking an infinitesimal distance and its corresponding infinitesimal interval of time, finding their ratio, and observing what happens when the time period in question is reduced until infinitely small (Feynman, Leighton, and Saads, 1963).

Leibniz's corresponding term for Newton's infinitely small increment, the fluent, was the *differential* (the term was later adopted universally). Leibniz was much less pragmatic than Newton. His idea of the calculus was closely connected to two of his broader philosophical themes: symbolism and continuity. Leibniz believed in the existence of simple and primitive ideas, like letters of the alphabet, that could be fitted together. He sought to devise a *universal technical language* similar to algebraic notation, a *universal or philosophical language* in which characters and words directly expressed logical connections between concepts, and even a *calculus ratiocinator* which resembled a formal system of logic that could immediately uncover errors and even eliminate them. All three projects related to an ideal of religious peace.

Leibniz's "principle of indiscernibles" was instead closely connected to the question of continuity in which two things in nature could never be exactly alike; that is, to the extent to which no internal or intrinsic differences can be detected. If two objects share the identical characteristics then they are the same thing. If they are different from one another they must display some differences, even if imperceptible or infinitesimal. The infinitesimal variations that defied expression in algebra could instead be expressed by the calculus, a method which required its own symbolism for integration (areas) and differentials (infinitesimal variations). Leibniz also formulated the rules for manipulating infinitesimal quantities.

Leibniz was no atomist and remained aloof from Cavalieri's indivisibles. He thought of infinitesimals as well-established inventions. Inventions because they did not correspond to a corpuscular reality, and well-established because they were justified not only by the calculus and the host of similarities between classical and modern geometry, but also by a metaphysical view of the world as a continuous hierarchy of infinites.

There was no shortage of controversy, criticism and incomprehension. For instance, in paragraph 130 of the famous *Principles of Human Knowledge* (1710), George Berkeley (1685–1753) attacked the "strange notions" that existed among his contemporaries: not only can finite lines be subdivided into an infinite number of parts, but each infinitesimal is divisible forever on into infinity. "In my days, mathematicians," he wrote many years later in a note to *Siris* (1744), "regardless of their claims of proof, embrace obscure notions and uncertain opinions and worry about them, contradicting one another and arguing like all men" (Berkeley, 1996: 650). The success of the calculus, he continued, proved absolutely nothing: the accuracy of results simply depended upon the fact that errors of lack and errors of excess compensated one another. Both

Newtonian and Leibnizian calculus introduced quantities that, depending on the case, were different from zero and equal to zero. Berkeley forcefully called attention to this "flaw" (Giusti, 1990: xlii).

The calculus proved to be an extraordinarily powerful tool. It paved the way for the study of dynamics, as well as the study of electricity, heat, light, and sound. By using infinitesimal calculus, science in the seventeenth and eighteenth centuries succeeded in solving, or practically solving, a vast number of problems in other fields. The same mathematical process that was used to calculate the rate of change in distance compared to time in an instant was used to calculate the *rate of change of one variable with respect to another* for a specific value of the latter. Such a process found its way not only into physics but also economics and genetics. According to Kline, "In order to treat the concept of instantaneous speed the mathematician has idealized space and time so that he can speak of something existing at an instant of time and at some point in space. He thereby obtains speed at an instant. The layman finds his imagination and intuition strained by the notions of instant, point, and speed at an instant and he might prefer to speak of speed during some very small interval of time. Yet mathematics produces through its idealization not merely a concept but a formula for speed at an instant that is precise and more readily applied than is the notion of average speed during some sufficiently small interval. The imagination may be strained but the intellect is aided" (Kline, 1953: 220).

16
Academies

The University

In the early Renaissance, Italian universities emphasized the study of law and medicine while in northern Europe theology and the liberal arts were more important. Italian students went to Oxford and Paris to study theology while other students crossed the Alps heading south to study medicine and law in Italy. Of the three major disciplines, law was the most prestigious, had the best paid faculty, and the highest enrollment. There were generally fewer professors and students of theology, though as a discipline it still exercised significant influence. In the faculty of medicine, a student could get a degree in "arts" or "philosophy" or else pursue a degree in medicine, sometimes called "arts and medicine" or "philosophy and medicine." The program was five years long and divided into two parts. The first two years were devoted to courses in logic (Aristotle's *Posterior Analytics*) and natural philosophy (including works like the *Physics*, *De anima*, *De generatione et corruptione*, *Parva naturalia*). In the second three-year part of the program, theory and practice were studied in the works of Hippocrates, Galen, and Avicenna. Arts education could include mathematics, humanistic subjects, and moral philosophy. Anatomy and surgery tended to be separate disciplines, as was botany. It was in the sixteenth century in fact that botany established itself as a completely separate field (Schmitt, 1975).

In university curricula, the teaching of mathematics occupied a decidedly secondary position. In the late sixteenth century, the University of Bologna averaged 22 professors of medicine. In 1590, the University of Pisa had 9, and there were 11 at the University of Padua in 1592. In most prominent universities, there was instead only one professor of mathematics for every dozen or so physicians. In the sixteenth century, moreover, "mathematics" included various areas of study including astrology, astronomy, optics, mechanics, and geography. As a result, a number of scientific disciplines gravitated toward the chair of mathematics. The term *cosmographia*, for example, first cropped up in Ferrara in the mid-sixteenth century and included geography and Ptolemaic astronomy. Many studies (see Schmitt, 1975: 47) have documented the frequent "incursions" of mathematicians into philosophy and the natural sciences. Meanwhile, theological studies increased dramatically following the Council of Trent. Prior to 1580, Bologna had only one chair in theology. There were three by 1550, six by 1600, and 9 by 1650 (Dallari, 1888–1924; Schmitt, 1975: 51).

We have already discussed Bacon and Descartes' criticisms of the universities. In England in particular, Bacon's critical attitude was responsible for important developments. The Puritan movement forcefully attacked both the inadequacies of the curricula and the backwardness of teaching methods. The attempt to introduce the new sciences into universities not only favored practical applications and "inventions" but also served to widen the circle of students. In the period between the outbreak of Civil War in 1642 and Cromwell's assumption of the title "Lord Protector" in 1654, a number of works critically addressed the question of what was being taught in universities (by John Milton, John Hall, John Dury). Even Thomas Hobbes complained in the *Leviathan* (1650) that university philosophy was simply Aristotelianism, that geometry was entirely neglected, and that physics was a lot of idle talk and no explanations.

A long struggle for independence, a decentralized government, and an international reputation as a tolerant and liberal country produced a strikingly different situation in the Netherlands. The population comprised a remarkable array of nationalities, and William of Orange was convinced that the key to national unity was the creation of a superior educational system, a policy endorsed by the Estates General. Universities were established at Leiden in 1575, Groningen in 1614, and Utrecht in 1636. The financial situation was good and high salaries attracted many foreign professors: during the entire seventeenth century 34 out of 52 professors at the University of Groningen were foreigners. Many students were foreign as well: between 1575 and 1835, 4,300 English-speaking students studied medicine at the University of Leiden. Though Cartesian philosophy was banned in 1656, traditional philosophies did not come to dominate as demonstrated by the swift acceptance of the anti-Aristotelian theories of Petrus Ramus (1515–72).

In the Netherlands, as in the rest of Europe, the university was not the only center for scientific study. Christiaan Huygens (1629–95) had studied at a university but then broke with that tradition. Antony van Leeuwenhoeck (1632–1723) was a textile merchant. Isaac Beeckman followed in his father's footsteps and was a candle merchant. Nor could one study any of the activities for which the Dutch were internationally famous – manufacturing precision instruments and machinery, shipbuilding, land reclamation, the building of canals and dikes – in any Dutch university (Hackmann, 1975).

That great season in European civilization that was Humanism did not have the same sort of dramatic effects on academic institutions as had the so-called "twelfth-century renaissance." As Westfall puts it: "in 1600, the universities gathered within their walls a group of highly trained intellectuals who were less apt to welcome the appearance of modern science than to regard it as a threat both to sound philosophy and to inspired religion" (Westfall, 1971: 106). It was the Scientific Revolution that produced actual alternatives to university culture, and created *different centers* for the construction and transmission of knowledge (Arnaldi, 1974: 14).

The Academy

The research institute is more closely associated with scientific pursuits than humanistic or literary ones. The aim of such an institute is not the spread but the advancement of knowledge, and the institute operates on the assumption that such advancement can be accomplished through the work of a group or team under the guidance of a director. The research institute was a nineteenth-century phenomenon, though there were of course precursors: for example, the observatory founded by Tycho Brahe (1546–1601) at Uraniborg in 1576 or the Paris Observatory directed by Gian Domenico Cassini (1625–1712).

The academies that were founded in the seventeenth century, even the most important of them, were not research institutes in the modern sense of the term. They did not aim at the transmission of knowledge but were instead places where information was exchanged, theories discussed, experiments analyzed and conducted, and, above all, where opinions and criticisms of the experiments and papers of members and external individuals were given. It is important not to project the characteristics of the later (and better known) scientific academies onto the academies of the sixteenth and seventeenth centuries. Nonetheless, a fundamental characteristic of even these early associations of scholars was their *rejection of solitary intellectual pursuit*.

By the term *academy*, wrote Girolamo Tiraboschi at the end of the eighteenth century, "I mean a society of erudite men governed by certain rules to which they are subject, that assembles together to discuss some erudite question; they produce and then submit to the scrutiny of their colleagues an essay of their own wit and fruit of their studies." Meetings, the setting of rules of conduct, and the critique of one another's work were three crucial elements. At the heart of any academy was the demand for collective work on a *collective subject*, and more importantly the need to submit intellectual production to the scrutiny of others and to public verification. The institution itself imposed its own rules: "its structure is that of a microsociety in imitation of real society." The academy co-opted its members through a sort of "rite of passage" which often assigned members a new name; it established itself as "neutral territory," complete with its own rules, within the greater, turbulent and agitated society (Quondam, 1981: 22–3).

The very name that many academies chose for themselves was, on the one hand, a way to express their method of research and their goals: the *Lincei* or the Lynxes, *Investiganti* or Researchers, *Cimento* or Experiment, *Traccia* or Evidence, *Spioni* or Spies, *Illuminati* or Illuminated. In other cases the name referred to the separation between the academy and society-at-large, revealing a climate of persecution or opposition that characterized some cultural settings: *Incogniti* (Disguised), *Secreti* (Secret), *Animosi* (Bold), *Affidati* (Trusted) and so forth (Quondam, 1981: 43; Ben David, 1971).

The First Academies

The first organization we can define (with as we shall see the necessary qualifications) as a scientific society was not the Academia Secretorum Naturae created in Naples by Giambattista Della Porta (1535–1615) but the Accademia dei Lincei (Academy of the Lynxes), established in 1603 by Duke Federigo Cesi (at the time 18 years old) and three young friends including the Dutch physician Joannes van Heeck. The first commitment the members made was to study together and give each other lessons. Cesi's family opposed his participation, so the group disbanded. The Academy was revived, however, in 1609. In 1610 Della Porta became a member as did Galileo in 1611.

Some scholars find the co-membership of two such different individuals, with utterly incommensurable worldviews, as a sign that the Academy was unclear about its program. Yet the climate of secrecy and the group's early "Paracelsian" orientation do not cancel out the significance of Cesi's goal "to read this great, true, universal book of the world" in order to "represent things as they are" and "explain how to alter and change them." It was Cesi's intention that a detailed constitution – the *Linceografo* – precisely govern admission into the Academy and the lives of its members. The text was never published and had few practical applications. The one rule, however, was always followed; namely the prohibition that a member of the Lincei belong to a religious order.

As we have said, the academies were micro-societies that operated within a greater articulated society. Like all scientific academies, the Lincei aspired to claim (in a limited sphere) the right to independent knowledge, maintaining the non-conflict between science and religion and science and society. Members of the Lincei, "by explicit virtue of their constitution, have outlawed all discussion of subjects not natural or mathematical and have deemed political discussions as unwelcome, and so to be left to others" (Olmi, 1981: 193). Elements that clearly marked the activities of the Lincei as "scientific" included their focus on mathematics and natural experiments, the quarrel with universities, the desire to clearly differentiate themselves from literary academies, the elevation of artisans (compared to "pompous and pedantic professors"), and the insistence on the "public" nature of knowledge. Though certainly the first Lincei went farther in planning out a research plan than in actually carrying it out. In the words of Cesi, the Lincean philosopher "will not limit himself to the writings or sayings of this or that teacher, but, in the universal exercise of contemplation and practice, will seek all knowledge that comes our way, whether of our own invention or communicated by others" (Altieri Biagi, 1969: 72).

The Accademia del Cimento, correctly described as a "plain product of court life" (Hall, 1963: 135), was a short-lived enterprise: 1657 to 1667. The group of professors, researchers, and artisans did not come together spontaneously but by design of the Medici prince Leopold, brother of Grand Duke Ferdinand II and a great admirer of Galileo. Leopold himself participated in the meetings of the academy, which counted among its members Vincenzo Viviani (1622–1703), Francesco Redi (1626–98), Nicolo Steno (1638–86), Alfonso Borelli (1608–79), Lorenzo Magalotti (1637–1712), and the Aristotelian Ferdinand Marsili.

When Leopold was appointed Cardinal in 1667 the meetings ceased, in part because of the disagreements among its members. Precisely because it was a product of the court, the Accademia del Cimento lacked both the structure and the features of a modern scientific institution. There were no rules or obligations for either the membership or the prince. The meetings were informal and not held in a fixed location. There was neither a budget nor a treasury, and the patrons and protectors of the academy viewed its activities as entirely celebratory ones (Galluzzi, 1981: 790–5). Leopold's cultural agenda certainly aimed at supporting and disseminating the new scientific ideas of which Galileo was the main and combative champion. Yet the rigidly experimental method adopted by the members tended to exclude conclusions of a theoretical character. If there are theoretical "speculations" in the *Saggi di naturali esperienze* (only published in 1667 and translated into English in 1684), then they "need to be understood as the concepts or ideas of particular academicians but not of the Academy as a whole which aims only to experiment and record" (Altieri Biagi, 1969: 626). The Accademia del Cimento and many other academies shared this voluntary restriction to "experimentalism." In Italy, that approach derived in part from the specific environment created following the condemnation of Galileo, though at the same time the Cimento proved to be an efficient instrument of Galilean vindication and propaganda (Galluzzi, 1981: 802–3).

The Neapolitan Accademia degli Investiganti (1663–70) took a different approach. For Tommaso Cornelio (1614–84), Leonardo di Capua (1617–95), and Francesco d'Andrea (1624–98), the reform of philosophy and science was inseparable from professional and civic renewal. The Neapolitan innovators tended to unite Galileanism and Cartesianism to Telesio and the tradition of Renaissance naturalism (Torrini, 1981: 847, 853, 876).

Today, the assertion of a direct continuity from the Lincei, the Cimento, and the Investiganti to the great European Academies no longer seems tenable (Galluzzi, 1981: 762). The stark differences in political and religious situations together with differences in philosophical traditions and discordant (often divergent) images of science created a complex intersection between the spontaneous association of scientists and the official political interest in such an enterprise.

Paris

Patronage operated in France as well, but real or "ideal" meetings between scientists also took place spontaneously as in the case of the complex network of relationships and correspondences maintained by Marin Mersenne (1588–1648) with almost 40 different scientists throughout his lifetime. All this at a time before the circulation of newspapers and journals, when written correspondence was still the primary conduit for intellectual exchange. In another example, between 1615 and 1662 the Cabinet des frères Dupuy was a center for scientific discussion. And an even more important assembly was the Academy of Montmor founded in 1654 by Habert de Montmor (1634–79); the meetings took place at his house and included a host of distinguished guests.

The 345 public "lectures" that were given every Monday afternoon at the Bureau d'Adresse in Paris between 1633 and 1642 were unusual. The Bureau was founded as a business and medical studio around 1630 by Théophraste Renaudot, a physician from Loudun, and gathered together a mix of dilettantes and curiosity-seekers, lawyers, physicians, and *beaux*

esprits. According to detailed records, the discussions were informal, spirited, and included a broad range of cultural topics. The dominant tendency in debates on philosophy, medicine, mathematics, astronomy, and
physics was that of a compromise between the old and the new. It was
the firm belief of the organizers, supported by Cardinal Richelieu, that
the advancement of learning required freedom of discussion during which
truths could be scrutinized and, under the weight of criticisms, modified
or abandoned. It was the feeling of many of the members that, contrary to the university approach, a theory was not an "invincible entity"
(Borselli, Poli, Rossi, 1983: 13, 32–6).

In 1663, Samuel Sorbière asked Jean-Baptiste Colbert (1619–83), Louis
XIV's minister of finance, if the State would help preserve and transform
the Montmor group. Thus was founded the Académie Royale des Sciences in 1666. Christiaan Huygens (one of the group's foreign members)
explained in a memo to Colbert his plans for experiments on vacuums,
gunpowder, wind resistance, and impact. The group's "primary and most
useful project" according to Huygens would be to "work on natural history along the lines of Bacon's plan." That great history "made up of
experimentation and observation is the only method for achieving the
knowledge of causes for all things perceptible in nature." It was necessary, he concluded, "to begin with those arguments deemed most useful
and assign many and various members to them who will report on a
weekly basis; in this way all will proceed in an orderly fashion and highly
significant results will be obtained" (Bertrand, 1869: 8–10).

The Académie was the first "research institute" to be directly supported by the state. The first academicians received salaries which ranged
from the 6,000 French livres annually to Gian Domenico Cassini to the
1,500–2,000 received by the French members; though given how slowly
one ascended through the ranks, the post of academician was not a particularly remunerative one. There were 16 initial academicians and the
number had grown to 50 by the late seventeenth century. In 1699 the
number had risen to 70, and the Académie introduced a strict hierarchy
of posts that remained in place until the French Revolution.

Colbert pursued precise goals: the planned growth and expansion of
industry, commerce, navigation, and military technology. But he was a
far-sighted politician and allowed the academicians great autonomy. The
Académie promoted some scientifically significant undertakings: Jean
Picard (1620–82) computed the radius of the Earth, for example, and
Jean Richer (1630–96) computed the distance from the Earth to the sun.
Though after the death of Colbert in 1683 achievements of a notably
more practical nature predominated, like the improvement and main-
tenance of the fountains in the royal gardens. Louis XIV, for his part,
considered the Académie to be an extra jewel in his crown and called the

academicians *mes fous* (my fools). After the revocation of the Edict of Nantes in 1695, the Académie lost its most illustrious foreign members such as Huygens and Roemer.

According to Roger Hahn, a spirit of research aimed at the rational comprehension of nature did not match the needs of the *ancien régime* French society. Many academicians were used as government consultants while others were forced out of economic necessity to accept positions as teachers and administrators. On this basis, the profession of scientist did not emerge as autonomous and viable, and the eighteenth-century scholar was subject to centrifugal forces that pulled him in other directions (Hahn, 1971).

London

The Royal Society of London officially predates the Paris Académie; the name was used for the first time in 1661 and on July 15, 1662 the Society was officially incorporated and approved by King Charles II. Its membership included all but one of those who belonged to the group that since 1645 had met at Gresham College and which had originally been founded in 1597 at the home of a wealthy merchant. According to the recollections of the mathematician John Wallis (1616–1703), penned more than thirty years later, the group met weekly in London; the members made contributions to cover the cost of experiments; "precluding matters of theology and politics [. . .] we discoursed of the circulation of the blood, the valves in the veins, the venae lactae, the lymphatick vessels, the Copernican hypothesis, the nature of comets and new stars, the satellites of Jupiter, the oval shape (as it then appeared) of Saturn [. . .] the weight of air, the possibility or impossibility of vacuities, and nature's abhorrence thereof, the Torricellian experiments in quicksilver" (Hall, 1963: 142; Johnson, 1957).

The new Society was a composite, bringing together threads of mathematical and astronomical traditions, those of medicine and chemistry, and a dose of the "technological" as well. In addition, Robert Boyle, one of the most influential members of the new institution, had been very interested in the project for an "Invisible College" (as we learn from his correspondence of 1646–7). This latter group revolved around German-born Samuel Hartlib, a disseminator of the "pansophism" of Amos Comenius (Johannes Amos Komenski, 1592–1670), and his work in England (after 1628). According to some scholars Boyle then constituted something of a link between the Hermeticism and utopianism that was so strong in Germany and the new experimental science (Rattansi, in Mathias, 1972: 1–32).

The only "royal" thing the Society possessed was its name. It received no money from the crown and was exclusively member-supported, hence its ultimately very large membership. The stipends of the secretary and the curator of experiments Robert Hooke (who for this reason has been described as the "first professional scientist in history") were relatively meager. The Society's first project was a typically Baconian one: the compilation of various "histories:" of mechanics, astronomy, trades, agriculture, navigation, cloth manufacturing and dyeing, etc. The ambition to do truly collective research was soon abandoned, though, unlike the case of many other similar groups, "when a paper was read or an idea discussed the matter was rarely dropped before some experiments had been performed before the assembled company" (Hall, 1963: 144). Furthermore, the recent scientific literature was examined carefully and any experiments described repeated by the Society. Hooke and Boyle were especially active members, as was the Society's secretary, Henry Oldenburg (1615?-77), a German who had established himself in England in 1653 and was at the center of a broad network of personal and epistolary contacts.

Unlike the Académie des Sciences, the Royal Society was wholly independent of the State: it was given the privilege of using the diplomatic postal service for its correspondence abroad and its only duty was running the Greenwich Royal Observatory (established in 1675). It had become an instrument "for the establishment of constant intellectual exchange between all civilized countries" and the Society intended to act as "the universal bank and free harbor of the world." Writing in 1667, Thomas Sprat explained that the Society had "freely admitted men of different religions, countries, and professions of life. [. . .] For they openly profess not to lay the foundation of an English, Scotch, Irish, Popish or Protestant philosophy, but a philosophy of mankind" (Sprat, 1966: 63).

Berlin

One does not find a true research institute in the Germanic countries till a bit later; certainly the Leopoldinisch–Carolinische Deutsche Akademie der Naturforscher, established in Schweienfurt in 1652 by four physicians who called it the *Accademia Naturae Curiosorum* (after the name used by Della Porta in the sixteenth century), does not qualify (Kraft, 1981: 448). Late seventeenth-century Germany was a mosaic of large and small states, some of which were Catholic and others Lutheran: from large Prussia–Brandenburg to duchies, cities, and autonomous villages. University reform followed a model devised by a disciple of Luther's, Philip Schwarzerd, also known as Melanchthon (1497–1560): one had first to graduate from a faculty of Arts and Philosophy before enrolling in the faculties of Law, Theology, or Medicine. Despite widespread poverty and many wars, Germany was a well-educated country. Compulsory education for children was already the rule in Prussia in the first decade of the eighteenth century (Farrar, 1975).

The great philosopher, mathematician, and historian Gottfried Wilhelm Leibniz (1646–1716) had a fairly low opinion of universities. He felt they were antiquated, isolated and almost completely ossified. He headed the project for a great Academy and addressed the difficult problem of financing. He fashioned his academy on the French model but rejected state control and insisted upon the need for great autonomy. He also felt it was the task of an academy to create a great encyclopedia of knowledge (Hammerstein, 1981: 413–18). What finally transpired did not correspond to Leibniz's original aspirations and he linked his project of an academy to those objectives considered less than noble by Bacon: the exaltation of one nation in the face of others (Hall, 1954). Leibniz believed that through the creation of an academy, the German nation and language would be enhanced, science would be enriched, industry and commerce would expand, and Universal Christianity would be spread through science.

The Societas Regia Scientiarum was established based on Leibniz's plan on July 11, 1700 under the sponsorship of the Elector (later King) of Brandenburg–Prussia, Frederick I. The academy was officially recognized on 19 January 1711. At the suggestion of Voltaire, it was reorganized by Frederick II who, in 1746, called on Pierre-Louis Moreau de Maupertuis (1698–1759) to direct it. It was renamed the Königliche Preussische Akademie der Wissenschaften (Royal Prussian Academy of Sciences). The leadership of Maupertuis represented the peak of French influence on German culture: French was adopted as the official language of the Academy and until 1830 the *Abhandlungen* were still entitled the *Mémoires*. The Berlin Academy had an operating theater, a botanical garden, and collections of natural histories and scientific instruments.

Bologna

Many of the scientific societies that flourished in Europe had two basic characteristics: (1) a shift from groups with broad interests to organizations that were specifically scientific; and (2) within those organizations the "experimentalists" gained a dominant position. In the final quarter of the seventeenth century, because of the impact of Cartesianism and mathematical and experimental neo-Cartesianism (represented by Huygens, Leibniz, and Malebranche), scientific organizations displayed a tendency toward professionalization: the Societies became centers for the discussion of *results* more than *ideas* (Hall, 1963).

The Institute of Science in Bologna showed just these tendencies. The Institute was not to be a place for lessons or scientific discussions, and "all its efforts were best expended toward the method of observation, experiments, and other such similar things" (Tega, 1986: 19). For Italy, the Bologna Academy of Science was a novelty. Bonaventura Cavalieri was active in Bologna between 1626 and 1647, and Marcello Malpighi from 1666 to 1691. In 1655 the first edition of Galileo's *Opere* was published there (which because of the censure was missing the *Dialogues* and the *Letter to Madame Cristina*). Furthermore, from 1690 the city was home to the Accademia degli Inquieti whose members were interested in astronomy, infinitesimal calculus, and life sciences. Luigi Ferdinando Marsili (1658–1730), who put his home and collections at the disposal of the group, fruitlessly attempted to reform the universities and in 1790 published his *Parallel between the University of Bologna and others beyond the mountains*.

Publications

It is pointless to even attempt to list the numerous newspapers, journals, magazine, collections, and periodicals that reported on the impressive volume of work that came out of academies and European scientific societies. Three in particular, however, merit special attention. The first strictly scientific European journal, the *Philosophical Transactions*, was founded in 1665 by Henry Oldenburg; it carried the imprimatur of the Royal Society and published its correspondence. The *Journal des Savants* began publication that same year in Paris and reported on mathematics and natural philosophy, as well as history, theology, and literature. Finally, the *Acta Eruditorum*, which reviewed books from every field, appeared in 1684 in Leipzig: the *Acta* were published in Latin and were accessible to scholars and scientists throughout Europe.

17
Newton

The *Principia*

The *Philosophiae naturalis principia mathematica*, or *Principia*, published in London in 1687, is a work that will never cease to amaze its readers. It is the synthesis of Newton's experimental and mathematical genius. It represents the ultimate expression and coherent organization, both methodologically and theoretically, of the revolution in science begun by Copernicus and Galileo. The work, so long in the making and so well-known, was destined not only to provide the essential elements of the scientific and philosophical credo of the eighteenth century, but also gave shape to a view of the world and the laws of nature that became the cultural legacy of the educated masses. For over two centuries – until the so-called "crisis in classical physics" – *all of physics* was essentially defined by Newton's view.

The very title of Newton's masterpiece expressed his stance toward Cartesian physics: the principles of philosophy are mathematical in nature. Unlike Descartes, Newton used mathematical language to express the principles of natural philosophy, while at the same time appropriating the culture of experimentalism and incorporating into his scientific method the Baconian mistrust of *hypotheses* that were not empirically based. Despite the fact that Newton had already invented the calculus some twenty years before writing the *Principia*, he chose instead to present his masterwork (with some exceptions) in the classical language of geometry. Newton so admired classical geometry that he regretted not having paid more attention to Euclid's *Elements* before applying himself to Descartes and the modern algebrists (Westfall, 1980: 98). However, as has been pointed out before, behind Newton's façade of classical geometry lay thought patterns characteristic of infinitesimal calculus (Whiteside, 1970; Westfall, 1980: 424).

Newton modeled the organization of the *Principia* after Euclid: the *definitions* of mass, force, and motion are followed by the *axioms* or laws of motion which are followed by a list of assumptions which he called *propositions* or *lemma*; and finally a number of *corollaries* and *scholia* (comments or explanatory notes). In chapter 15 we briefly mentioned the dispute between Newton and Leibniz over who had invented the calculus. How ironic then that eighteenth-century Newtonians would expound the new physics of the *Principia* and expand the applications of Newtonian mechanics by using Leibniz's version of the calculus.

Newton's physics differed from that of Descartes in more than just expository technique and method. Koyré has pointed out that the Newtonian world in contradistinction to the Cartesian one was composed not of two elements (extension and motion) but three: *matter*, an infinite number of mutually separated and isolated particles, hard and unchangeable but not identical; *motion*, that strange and paradoxical relation-state that does not affect the particles in their being but transports them hither and thither in the infinite, homogeneous void; and *space*, or that very infinite and homogeneous void in which, unopposed, the corpuscles (and bodies built of them) perform their motions (Koyré, 1965: 13).

The *Principia* begins with the "Definitions." The first defines the quantity of matter or *mass* of a body as the product of density and volume and clearly distinguishes a body's mass (constant irregardless of location in the universe) from its *weight* which depends on the force of gravity and therefore varies with distance. Newton did not consider weight an absolute value. In Book III he equates gravity with centripetal force: the force of attraction exerted by a body is proportional to its mass and the weight of an object of equal mass is different on the surface of different planets. The second definition uses the expression *quantity of motion* (momentum) to mean the product of a body's mass and its velocity. The third definition refers to the *inherent or innate force* of matter by which every body continues in its present state, whether at rest or in uniform motion along a straight line: "upon which account this *vis insita* may, by a most significant name, be called "inertia" or "force of inactivity." According to the fourth definition, an *impressed* force is an action exerted on a body in order to change its state of rest or uniform rectilinear motion. The term *centripetal force*, or that which "seeks the center," (for example, the force that keeps the planets in their orbits), was coined by Newton and is used in the fifth definition which states that bodies tend toward a center. Its opposite is *centrifugal force* (coined by Huygens) which is experienced by bodies as they move away from a center.

The "Scholium" is a discussion of space, time, and motion. The perfectly equivalent states of rest or uniform rectilinear motion can be determined only in relation to other bodies either at rest or in motion. Because the available frames of reference are limitless, Newton believed the eternal and uniform flow of time (*absolute time*) and the infinite extension of space (*absolute space*) are the coordinates by which one must ultimately define a body's state of rest or motion. Relative space and relative time are quantities conceived in relation to sensible objects whereas the senses must be disengaged in philosophy: "Absolute, true, and mathematical time, of itself and from its own nature, flows equably without relation to anything external, and by another name is called 'duration' [. . .]. Absolute space, in its own nature, without relation to anything external, remains always similar and immovable" (Newton, 1965: 109–10, 104–17; 1953: 17–18).

The Newtonian concept of the relationship between relative and absolute motion (a concept that remained firmly entrenched in the twentieth century) was expressed in his bucket experiment. A bucket full of water is hung by a long cord. The cord is twisted tightly and then let go. When a concave figure forms on the surface of the water it is observed that the water performs its revolutions "in the same times" as the bucket. In this case the water is in a state of relative rest in the bucket. The ascent of the water up the sides of the bucket, however, shows how hard it tries to recede from the axis of its motion and this force is the measure "of the true and absolute circular motion of the water."

Book I of the *Principia* begins by stating the three *Axioms or Laws of Motion* : (1) every body continues in its state of rest or of uniform rectilinear motion unless it is compelled to change that state by forces impressed upon it; (2) the change of motion is proportional to the motive force impressed and is made in the direction of the straight line in which that force is impressed; and (3) to every action there is always an opposite and equal reaction, or the mutual actions of two bodies upon each other are always equal and directed to contrary parts: "Whatever draws or presses another is as much drawn or pressed by that other: If you press a stone with your finger, the finger is also pressed by the stone" (Newton, 1965: 117–20; 1953: 26). Some of the theorems and corollaries that Newton deduced from these laws and the prefatory definitions include, for example, the theorem of the composition or parallelogram of motions: a body acted on by two forces simultaneously will describe the diagonal of a parallelogram in the same time as it would describe the sides by those forces separately. He also deduced Kepler's three laws of planetary motion from the laws of dynamics. Kepler's law of areas applies when a central force causes a body to deviate from its inertial direction. When centripetal force varies inversely as the square of the distance, a body will, depending on its tangential velocity, follow a "conic" orbit: an ellipse, parabola, or hyperbola.

Book II of the *Principia* turns from the problems of point masses moving without friction to the problem of bodies moving through resisting fluids. These pages represent pioneering work in the mechanics of fluids and essentially constitute the beginning of fluid dynamics. Book II also represents the demolition of Descartes' theory of vortices. Newton demonstrated that the motion of a vortex is not self-sustaining: it continues in uniform motion only as long as an external force continues to turn its central body. What is more, that motion will inevitably slow down as the energy it disperses into space is "swallowed by space." A vortex can never yield a planetary system according to Kepler's laws: "the theory of vortices runs completely counter to astronomical appearances and does more to obscure than explain celestial phenomena" (Newton, 1965: 593).

Book III is a presentation of "the organization of the world system." Newton felt it was necessary to state some "rules of reasoning in philosophy" in order to move from the definitions, axioms, theorems, and proofs to a description of the world.

Rule 1 states that "we are to admit no more causes of natural things than such as are both true and sufficient to explain their appearances." The rule affirms the *simplicity* of nature which "does nothing in vain" and "does not abound in superfluous causes." With this rule Newton introduced "Ockham's razor" to modern science: *Entia non sunt multiplicanda praeter necessitatem* (do not multiply entities beyond that which is necessary) or *Frustra fit per plura quod fieri potest per pauciora* (one vainly does with many things what can be done with few). The statements in this form – not original to the writings of William of Ockham (d. 1347) – were used by the schools of empiricism and nominalism to formulate the principle of economy or simplicity.

Rule 2 states that "therefore to the same natural effects we must, as far as possible, assign the same causes." The rule asserts the *uniformity* of nature or the general validity of natural laws: the causes of respiration are the same in man as in animals; stones fall in exactly the same way in Europe and in America; and light is reflected the same on Earth as on other planets.

Rule 3 states that "the qualities of bodies, which admit neither intensification nor remission of degrees, and which are found to belong to all bodies within the reach of our experiments, are to be esteemed the universal qualities of all bodies whatsoever." Here the *homogeneity* of nature is declared: its entities are invariable, regular, and predictable. The evidence of experiments is not to be relinquished "for the sake of dreams and vain fictions of our own devising; nor are we to recede from the analogy of Nature, which is wont to be simple and always consonant to itself." The qualities of bodies "are known only by our senses, so all those that in general agree with sensation can be considered universal."

The conclusions we reach by induction are valid when the senses are employed: for example," [t]hat all bodies are impenetrable, we gather not from reason but from sensation. The bodies which we handle we find impenetrable, and thence conclude impenetrability to be a universal property of all bodies whatsoever." The inference, however, reaches even beyond the realm of the senses, and Newton continues by saying that "hence we conclude the least particles of all bodies to be also all extended, and hard and impenetrable, and movable, and endowed with their proper inertia. And this is the foundation of all philosophy."

In Rule 4 Newton states that "in experimental philosophy we are to look upon propositions inferred by general induction from phenomena as accurately or very nearly true, notwithstanding any contrary hypotheses that may be imagined, till such time as other phenomena occur by which they may either be made more accurate or liable to exceptions." This rule states the necessity of *confirming* theories. Newton adds that from "this rule must follow that the argument of induction may not be evaded by hypotheses." Scientific theories must agree with experiments and be considered true as long as the accord exists (Newton, 1965: 609–13; 1953: 3–4).

The Rules are followed by a description of the system of the world in which Newton demonstrates that the orbits of the satellites around Jupiter and Saturn, and the orbit of Earth and the other planets around the sun, all obey Kepler's planetary laws. He calculates the Earth's mass and shows that the precession of the equinoxes are due to the shape of the Earth and the inclination of its axis, which at times depends on the combined effect of the attraction exerted by the moon and sun. This combination of forces acting upon the Earth was also sufficient to explain the tides. Newton was furthermore able to demonstrate that comets, the sudden and inexplicable appearance of which had for thousands of years seemed to defy the perfection of celestial motion, are governed by the same dynamic laws that govern the solar system. The comet of 1681 moved in a parabola (as predicted by Kepler's first law) and swept through (according to the second law) equal areas in equal times.

Newton states the law of *universal gravitation* in Book 3: in the universe, one body attracts another with a force that is directly proportional to the product of their masses and inversely proportional to the square of the distance between them.

$$F = G \left(m_1 m_2 / D^2 \right)$$

Where F is the force of attraction, m_1 and m_2 are the two masses and D is the distance between them. G is a constant *factor*: the value remains the same whether the mutual attraction is between Earth and an apple, Earth and the moon, the sun and Jupiter, or between the stars.

So Newton was able to formulate a single law to simultaneously explain the fall of an apple to the ground, the movement of the planets around the sun, and the phenomenon of tides. Many regarded the calculation in Book III in which it is shown that the moon is kept in its orbit by the same gravitational force that causes bodies to fall to the ground on Earth as one of the central points of the *Principia*. The "centripetal force that holds the different planets in their orbits is the very same gravitational force." Newton was greatly moved by his discovery that a single force keeps the planets orbiting around the sun, the satellites of planets in their orbits, causes bodies to fall on Earth, and is responsible for the tides. The outcome was a single worldview and the definitive union of terrestrial and celestial physics. The long-held belief in the essential difference between the heavens and Earth, and mechanics and astronomy, had been shattered – as was the "myth of circularity" that had influenced the course of physics for over a thousand years and weighed so heavily upon Galileo's work.

The "General Scholium"

The second edition of the *Principia* ends with a "General Scholium" addressing the perfection of planetary motion. Newton felt that "mere mechanical causes" could not have been the source of order in the cosmos, and since the existence of the world does not depend on the principles of mechanics, one must ultimately appeal to final causes. The sheer variety of objects in the world cannot have sprung from a blind metaphysical necessity and blind fate could never keep all the planets moving in the same direction in concentric orbits. Order in the cosmos is the result of a choice. "This most beautiful system of the sun, planets, and comets could only proceed from the counsel and dominion of an intelligent and powerful Being." He who ordered the universe placed the stars at great distances from one another "lest the system of the fixed stars should, by their gravity, fall on each other" (Newton, 1965: 792–3; 1953: 42). Similarly, the eyes, ears, brain, heart, wings, and instincts of animals and insects must be the product of the wisdom and skill of a powerful and eternal agent (Newton, 1779–85: IV, 262). All of space is the *sensorium* of Newton's personal and transcendent God. He "governs not as the soul of the world but as Lord over all; *Lord God* or *Universal Ruler*." He is omnipresent and "as the blind man has no idea of colors, so have we no idea of the manner of which the all-wise God perceives and understands all things" (ibid.: 794, Newton: 1953: 44).

The final portion of the "Scholium" revisits the topic of gravity. Newton explains that he had used the force of gravity to explain the phenomena of the heavens and the seas but had not yet shown the cause of that force itself. His famous statement about hypotheses follows: "But hitherto I have not been able to discover the cause of those properties of gravity from phenomena, and I feign no hypotheses; for whatever is not deduced from the phenomena is to be called a hypothesis, and hypotheses, whether metaphysical or physical, whether of occult qualities or mechanical, have no place in experimental philosophy. In this philosophy particular propositions are inferred from the phenomena and afterward rendered general by induction. Thus it was that the impenetrability, the mobility, and the impulsive force of bodies, and the laws of motions and of gravitation, were discovered. And to us it is enough that gravity does really exist and act according to the laws which we have explained, and abundantly serves to account for all the motions of the celestial bodies and of our sea" (ibid.: 794; Newton: 1953: 45).

Cartesian physics, and the mechanical philosophy in general, tended to view all phenomena as movements themselves part of a known model (impact, pressure, etc.). Newtonian physics used the principle of "action at a distance," which did not appear immediately compatible with a mechanical model. Cartesians throughout Europe and Leibniz himself felt that Newton had reintroduced the "occult properties" of Scholasticism into the new science of physics, something from which it had long struggled to free itself. They felt that Newton had abandoned the solid ground in which the new physics had been able to take root and thrive. This controversy was destined to be a long-lasting one. Though many eighteenth-century materialists explicitly referred to the rigid mechanicism of Descartes, the interlacing of mechanics and deism found in Newtonian philosophy ultimately came to dominate European Enlightenment culture.

It is important to point out that until the middle of the eighteenth century *two schools of physics* co-existed. In Voltaire's *Philosophical Letters* (1734), he not only compared English tolerance and freedom to the still feudal French regime but also Newtonian physics to Cartesian physics: in Paris the world is shaped like a lemon but in England it is shaped like a turnip. "A Frenchman who arrives in London finds that many things are different in natural philosophy, like in everything else. He leaves behind a full world only to find an empty one. In Paris we know the universe is made up of subtle matter but in London one knows no such thing. In France we believe the pressure of the moon causes the tides; the English believe the sea gravitates toward the moon [. . .]. According to the Cartesians, everything occurs as a result of an incomprehensible impulse yet according to Newton it occurs because of a force of attraction, the cause of which is not known" (Voltaire, 1962: I, 52).

The *Opticks*

Opticks, Treatise of the Reflexions, Inflexions and Colours of Light was published in London in 1704 (Newton was 62 years old) and reissued twice (in 1717 and 1721) in his lifetime. Newton himself supervised the 1706 Latin translation. Each edition included significant changes and Newton reworked some of the problems which he had already dealt with amply at the end of the 1660s and over the course of the 1690s. The *Opticks*, like the *Principia*, is divided into three books. Book I begins with a series of definitions and a group of axioms that establish the general principles of optics. These are followed by the propositions and theorems which explain, *more geometrico*, the experiments pertaining to geometrical optics, the doctrine of the composition and dispersion of white light, the aberration of lenses, the rainbow, and the classification of colors. Book II discusses questions relating to color, rings of interference, and the phenomena of light interference in thin plates. Book III describes a series of experiments investigating diffraction and the colored fringes that are produced in the presence of minute obstacles and sharp plates.

Robert Hooke revisited the Cartesian theory of the nature of light in *Micrographia* (1665). In a full mechanical universe, light, like sound, spreads in waves and Hooke described the laws of refraction and explained light as the result of vibrations in the aether. Newton's work on light and color relied on Kepler's *Dioptrice*, Descartes' *Dioptrique* (1664) in Latin, Francesco Maria Grimaldi's *Physico-mathesis de limine, coloribus et iride* (1665), Boyle's *New Experiments* (1667), and Isaac Barrow's *Lectiones opticae*, to which Newton himself had contributed.

Newton's position on the nature of light – waves or particles – was complex and surely in part related to a heated dispute with Hooke that took place between 1672 and 1676. Some scholars, he felt, were inclined to believe that light was composed of multitudes of unimaginably small and swift corpuscles emanated by bodies. Others, such as Grimaldi and Huygens, felt light was a movement in a medium. Grimaldi believed that light was like a fluid through which waves moved and Huygens suggested that longitudinal waves moved through a stationary fluid. Newton did not intend to get involved in any arguments he considered useless. He never outright confirmed the corpuscular theory though he made ample use of it. His theories were founded on experimental data and the solutions he proposed in which light was either corpuscular or wave-like in nature depended on the particular case at hand, although he was certain that the wave-theory could not explain either the rectilinear spread of light or the formation of shadows behind obstacles. The clash between supporters of the two theories worsened at the end of the seventeenth century into what was essentially a contrast between two schools. The division led to a radical contest between scientific metaphysics, which ended temporarily in favor of the corpuscular school in the eighteenth century, the wave-theory school in the nineteenth century, and today is represented by the post-1905 "complementary" quantum approach to optics (Bevilacqua and Ianniello, 1982: 245, 254).

In a letter to Henry Oldenburg, Secretary of the Royal Society, dated January 18, 1672, Newton explained that his theory of light was the greatest if not the most important discovery ever made in the study of nature (Newton, 1959–77: I, 82–3). Till then the numerous and often muddled descriptions of the nature of colors identified them as qualities inherent in the bodies upon which light acts but not of light itself. According to Aristotle, color was an inherent quality of bodies or produced by a blending of dark and light: red was white light blended with just a bit of dark and blue was white light to which the greatest possible darkness had been added. Paracelsus had believed colors were a manifestation of the Sulfurous principle; Descartes thought they were caused by different rates of rotation and translation of aether particles; and Hooke believed colors arose from differently inclined pulses. Newton solidly broke with both tradition and the positions of his contemporaries: he believed that the modification of light, which caused colors, was "an innate property of light." Colors are not generated by the reflection or refraction of natural bodies (as was generally believed) but "are original and innate properties, different for the different rays: some exhibit the color red and no other, others yellow and no other, still others green and no other, and so on for the rest" (Newton, 1978: 208). The question of color was no longer exclusively of perception: angles of refraction could be calculated; color was a physical problem – separate from the "psychological" problem – and could be dealt with mathematically. The color of a body is associated with how absorbent its surface is: "I shall call a ray that appears red, or makes objects appear red, or a producer of red [. . .] and so forth. Rays are not in fact colored. There is nothing in them but a certain power or disposition to simulate a sensation of this or that color. As the sound of a bell [. . .] is nothing more than a vibration and there is nothing in the air but a movement propagated by the object, and in the sensorium it becomes a sensation of the motion in the form of sound, so is the color of an object nothing more than a inclination to reflect this or that type of ray more copiously than others; in rays this is nothing more than their disposition to propagate this or that motion in the sensorium, and in the sensorium they become sensations of those motions in the form of colors" (ibid.: 393–4). Only in the early nineteenth century does the question of perception or psycho-physiology return to the study of optics and colorimetry. One of the greatest physicists of the twentieth century observed that "the phenomenon of color depends in part on the physical world. But also, of course, it depends on the eye, or what happens behind the eye in the brain" (Feynman, Leighton, and Saads, 1963: I, 2, 35–1).

The famous and complex prism experiment demonstrated that light "consists of differently refrangible rays" that are projected onto different points of a wall according to their degree of refrangibility: each grade of refrangibility corresponds to a basic color. Purple was the most refrangible and red the least. Colors do not exist as a result of disturbances of light; white light is not pure light but made up of rays with different characteristics and is the result of a blend of the colors in the "spectrum." White is neither an active color nor an "innate quality" of light but a sensation. The components of light can be separated and recomposed.

Newton's work on optics had important practical and technological results as well. When a colored fringe or the chromatic aberration of the lenses disturbed the images received through a telescope Newton decided to manufacture a reflecting telescope (or concave-mirror telescope) with a lateral eyepiece that received rays of light reflected by a prism. The mirror (made from an alloy of his own invention) was 25 mm. in diameter and the telescope itself only 15 cm. long; yet it magnified nearly forty times, which was much greater than the usual 180 cm. telescope. In 1671 Newton sent his telescope to the Royal Society. Early in 1672 he sent them a preliminary account of his theory of colors which appeared in *Philosophical Transactions* on February 19, 1672. With this publication, writes Westfall, "[s]wept along by the success of his telescope, Newton stepped publicly into the community of natural philosophers to which he had hitherto belonged in secret" (Westfall, 1980: 237).

Newton's Life

Isaac Newton was born in the small Lincolnshire village of Woolsthorpe on December 25, 1642, the same year that Galileo died. His father died when he was a year old, and he was sent to live with his grandmother when his mother remarried shortly thereafter. At age 12 he started grammar school in nearby Grantham. The boy who constructed such clever mechanical toys and filled the house with handmade sundials passed an unhappy childhood, particularly after his mother remarried. For instance, a 1662 list of sins reveals that he "threatened to burn my mother and father alive and their whole house with them." In 1661 he was admitted as a "subsizar" to Trinity College, a renowned community of over 400 at

Cambridge University. A subsizar was a poor student who earned his way through school by performing menial tasks for the instructors; in short, a servant, which is exactly what the same job was called at Oxford. Some of a subsizar's duties included waking the other fellows of the college, cleaning their boots, emptying their chamber-pots and so forth (Westfall, 1980: 57, 71). By 1664 he was no longer a subsizar and turned to the full-time pursuit of his studies. He earned his bachelor of arts in 1665, became a "junior fellow" in 1666, and earned a master of arts and became a "senior fellow" in 1668. The next year Isaac Barrow resigned as Lucasian professor of mathematics so that Newton could fill the position, which he in turn kept until 1704. Yet the 28 years he spent at Trinity College coincided with the most disastrous period in the history of both the college and Cambridge University. This surely had some bearing on Newton's lack of sociability and the solitude in which he lived (ibid.: 189, 190).

While at Cambridge, Newton not only read the Peripatetics, but Kepler's optics and astronomy, Descartes' *Geometrie*, Galileo's *Dialogues*, and the writings of Boyle, Hobbes, Glanvill and John Wallis, the mathematician. He became close friends with the theologian and Platonic philosopher, Henry More. He returned to live with his mother at Woolsthorpe during the plague of 1665–6. Those two or three years were remarkably, almost unbelievably, productive ones. He put to use a century's worth of learning and, on his own, devised a program of study that put him in the forefront of European science. Looking back on that period Newton reflected that he had spent more time on mathematics and philosophy then than at any other time of his life. By the end of 1665, at the age of 23, Newton had already formulated the binomial rule, found the direct method of fluxions (infinitesimal calculus), and deduced the nature of gravity: "the force which keeps the planets in their orbits must be reciprocally as the squares of the distances from the centers about which they revolve" (ibid.: 143–4).

Few were aware of Newton's discoveries because they remained unpublished. After succeeding Barrows as Lucasian professor, he gave a series of lectures on optics (the *Lectiones Opticae*). However, as a result of his dispute with Hooke (begun when he sent the Royal Society the letter on the nature of light and color), he decided against publishing them. Next followed a period devoted to the study of alchemy, theology, and the interpretation of the Apocalypse. He outlined the basic principles of celestial mechanics in *De motu corporum in gyrum*, and then turned to the writing of the *Principia* which was published when he was 45 years old.

The creative phase of his scientific research actually ends with the *Principia*. As we know, the *Opticks* – published in 1704 after the death of Hooke – was composed of previously written work. Furthermore, the appendix explaining the method of fluxions was also the result of work he had done over thirty years earlier. A review in the 1705 *Acta Eruditorum* sparked a long and nasty quarrel with Leibniz over who was the rightful inventor of the calculus and remains to this day one of the most famous controversies in the history of science (Hall, 1980).

Newton lived a bookish life in Cambridge and London until 1688 and the "Glorious Revolution." In 1689 he embarked on a career as a public servant and was appointed Warden of the Mint of London in 1696, a post he retained for thirty years. From 1689–90 he represented the Whig party in Parliament. In 1703 he was elected president of the Royal Society and exercised a tremendous influence on European intellectual life. The prestigious scientific society became something of a personal fiefdom. He died in 1727 at the age of eighty-five.

Newton had always been so absorbed in his work that it was not uncommon for him to spend entire nights at his desk. Engrossed in solving a problem, he often neglected to eat meals, leaving his untouched plate to be licked clean by his increasingly stout cat. He spent his life concealing deep religious convictions from others, and tended to be stand-offish in his relations with others. He had a "built-in censor and lived ever under the Taskmaster's eye" (Manuel, 1974: 16). His first and last romantic tie to a woman dates to his Grantham grammar school years. Humphrey Newton, his clerk at Cambridge for five years, reported that he had seen him laugh but once. The scathing and arrogant remarks made in letters to Hooke and Huygens reveal the frequency with which he lost his temper. He used John Keill and a committee appointed by the Royal Society as a smokescreen during his dispute with Leibniz (though Leibniz too preferred to remain anonymous). As Westfall has pointed out, it was as if Newton was consumed by childhood neuroses and stress from his work: he was a tormented and neurotic man who, in his later years, lived constantly on the verge of a nervous breakdown (Westfall, 1980).

A Brief Interlude on the Manuscripts

Before we discuss the "Queries" that close the *Opticks*, I would like to clarify a point that has had a bearing on the discussion thus far, and for which I naturally make no claim of originality. I am well aware that few of the many and well-trained Newton scholars of today would ever begin a study of Newton's natural philosophy with an analysis of his major works. I have chosen this unorthodox approach for two reasons. First, Newton's reputation and celebrity in the eighteenth, nineteenth, and early twentieth centuries was almost exclusively associated with his two great masterworks. The admirable, painstaking, and sophisticated research of outstanding contemporary scholars has unsettled a seemingly saturated field and radically transformed the significance of Newton as well as his place in history, yet it also has jeopardized the communication of such an *obvious* point to today's average reader. Which leads to the second reason. The great interest in his lesser known or even unknown works may lead to the paradoxical result that Newton be exclusively discussed, even in textbooks on the history of science or philosophy, in terms of works which he, for reasons of prudence or an extreme regard for privacy or both, decided to leave unpublished and therefore unknown to his public. I came to this decision the day I read a chapter about Newton in a textbook that cited *only* his unpublished works.

On Newton's death, the Royal Society refused to acquire his religious writings and returned them to his family, recommending they show them to no one. When Samuel Horsley, editor of Newton's *Opera omnia* (published between 1779 and 1785), discovered the papers in question, he reportedly, "in shock, slammed shut the lid of the trunk that held them." The renowned British economist John Maynard Keynes acquired some of Newton's manuscripts, and once he realized the great extent to which they dealt with alchemy, he made the provocative pronouncement that Newton was not the first great modern scientist but the "last of the magicians." Though the manuscripts dealt extensively with problems of mathematics, physics, optics, and "science," a good many were devoted to alchemy, Biblical chronology, the interpretation of the Scriptures and theological disputes, the Apocalypse, and the Hermetic notion of a secret, primitive knowledge. A number of institutions would not even bid on his papers: Cambridge University (which chose only some scientific papers), the British Museum, Harvard University, Yale University, and Princeton University. Israel, the recipient of a large part of the manuscript collection in 1951, only gave them to the University Library in Jerusalem 18 years after the fact (Mamiani in Newton, 1994: vi–vii).

Newtonian scholars are in general agreement that work published prior to 1945–50 – though some of it is still of fundamental importance – has in some ways been rendered obsolete by later studies using manuscript sources. Newton's unpublished mathematical and scientific papers became available only in the 1960s and 1970s (Newton, 1984; Herivel, 1965), as did his unpublished work on optics and philosophy (Newton, 1984; 1983b) and his correspondence (Newton, 1959–77). Furthermore, it has only been in the last two decades that the so-called *Scolii classici*, planned as an appendix to the second edition of the *Principia* and the *Treatise on the Apocalypse*, have seen the light of day (Newton, 1983a; 1991; 1994). One might say that scholars in the postwar period have been inundated with material, and even if they wanted to restrict themselves to the most essential work, this would mean roughly twenty volumes, with still more to come.

From this perspective, Newton has certainly met with a curious fate. Nothing like it happened to Copernicus, Descartes, Galileo, or much later, Darwin. The positivist portrayal of these scientists differs markedly from contemporary portrayals. However, it is one thing to discover a new manuscript, another to account for changes and advances in historical scholarship, and quite another for a mountain of manuscripts to suddenly appear after almost two hundred years. The view of Newton as a "positivist scientist" (still current today) was constructed by historians and scientists in the late eighteenth and nineteenth centuries, but has been contributed to by the persistent and tenacious refusal to even consider a vast quantity of work that revealed an unknown dimension to a seemingly familiar figure; a "familiarity" which, in this case, is related to the collective portrait that modern or positivist scientists have of themselves.

I have actually had two convergent goals in this brief section. The first has been to make the general reader aware of the importance of having opened the trunk containing Newton's notebooks and having studied them. The second has been to humbly remind somewhat overzealous specialists that had Newton left behind only his unpublished "scientific" writings, not to mention the rest of the contents of the trunk, it would have been absurd to close a book on the Scientific Revolution with a chapter devoted to Isaac Newton.

The "Queries"

This "unknown dimension" of Newton which I have just mentioned could be partially discerned in the final chapter of the *Opticks*, the question and answer section he called the "Queries." There were sixteen "Queries" in the first edition of the *Opticks*, 23 in the 1706 Latin translation, and 31 in the 1717 English edition. Newton tackles a vast number of problems in the last few "Queries:" the void; the atomic composition of matter; the electrical nature of the forces that hold atoms together; the polarization of light; occult properties; the insufficiency of mechanical causes; Cartesian metaphysics; the relationship of God to the world; the nature of God; the relationship between natural and moral philosophy; nature's ability to transform itself in different and strange ways; and experiments in alchemy.

In *Query 31*, the most well-known and discussed of all, Newton suggests that the macrocosm and the microcosm are governed by the same dynamic laws. The power of attraction between particles is very strong.

Newton rejects the beliefs of other philosophers – atoms with hooks, atoms glued together by the state of rest, which is an occult quality, or by nothing, or even cohesion by conspiring motions or relative rest amongst themselves. Newton instead states that he "had rather infer from their cohesion that their particles attract one another by some force, which in immediate contact is exceeding strong, at small distances performs the chemical operations above-mentioned, and reaches not far from the particles with any sensible effect" (Newton, 1953: 168; 1978: 591–2). The force that holds particles together is either gravity or very similar to it: "As gravity makes the sea flow round the denser and weightier parts of the globe of the earth, so the attraction may make the watery acid flow round the denser and compacter particles of Earth for composing the particles of salt" (Newton, 1953: 166; 1978: 589).

By virtue of the notion that physical and chemical properties are determined by the particulate structure of matter, it was now feasible to unite physics and chemistry into one greater scientific discourse. The smallest particles of matter are held together by a very strong attraction and form bigger particles whose attraction is weaker. Many of these bigger particles can combine to form even bigger particles whose attraction is even weaker, "and so on for divers successions, until the progression ends in the biggest particles on which the operations in chemistry depend" (Newton, 1953: 172; 1978: 598).

Thus the world is "conformable to herself and very simple" given that the great motions of the heavenly bodies are produced by universal gravitation and "the small ones of their particles" are produced by "some other attractive and repelling powers which intercede the particles." Why does motion in the world exist? The impact of very hard or very soft bodies voids their motion. In the case of a meeting between two elastic bodies, the elasticity produces a new motion that is nevertheless inferior to the initial motion. The force of inertia is a passive principle: "By this principle alone there never could have been any motion in the world. Some other principle was necessary for putting bodies into motion; and now they are in motion, some other principle is necessary for conserving the motion" (Newton, 1953: 174; 1978: 598).

In addition to the "passive principle" of inertia, there are also active principles in nature such as the cause of gravity, fermentation, and the cohesion of bodies. There are significant differences between a Newtonian God and a Baconian or Galilean one. Newton's God *is part and parcel of* his physics.

The Cosmic Cycles

Newton's active principles in some way explain the existence of motion in the universe: "it were not for these principles, the bodies of the Earth, planets, comets and sun, and all things in them, would grow cold and freeze, and become inactive masses; and all putrefaction, generation, vegetation, and life would cease, and the planets and comets would not remain in their orbs" (Newton, 1953: 175; 1978: 600). Because the universe tends toward decay and destruction, divine intervention is required for its preservation. He who has ordered the universe, as we know, has also established the "primitive and regular" position of the celestial orbits. The worthy placement of the sun, planets, and the comets "can only be the work of an omnipotent and intelligent Agent." The universe cannot have arisen out of a chaos by the mere laws of nature, although once order had been introduced by the Creator, "being once formed, it may continue by those laws for many ages." There are, however, "inconsiderable irregularities" in the system "which may have arisen from the mutual actions of the planets and comets upon one another and which will be apt to increase till this system wants reformation" (Newton, 1953: 177; 1779–85; III, 171–2; 1721: 377–8; 1978: 602).

Leibniz felt that Newton's God – who could create a world that would last for the ages but *not* forever and which required some correction from time to time – was a poor watchmaker indeed. The Newtonian world-machine is like a watch that moves imperfectly, runs down if left on its own, and needs God to wind it up again or repair it every now and then: "Sir Isaac Newton, and his followers, have also a very odd opinion concerning the work of God. According to their doctrine, God Almighty wants to wind up his watch from time to time: otherwise it would cease to move. He had not, it seems, sufficient foresight to make it a perpetual motion" (Leibniz–Clarke, 1956: 11).

The ardent Newtonian, Samuel Clarke, responded that even though the active force is continually and naturally diminished in the physical world and needs new energy, this did not mean the universe was flawed. Rather, it was purely due to the nature of matter, which is lifeless, inert and inactive. Newton's universe occasionally needed to be recreated or reorganized. Newton's cosmogony and this theme of the "reorganization" of the universe did not receive much attention from scholars of Newton until the last few decades. Newton had always been presented as the exponent of a mechanistic science, the primary model of which was an absolutely static world, a world explained in the classic (and certainly fundamental) terms of the difference between relative and absolute time. Yet there are even more subtle and nuanced interpretations within this framework. For instance, there has been ample documentation recently of the role thirteenth- and fourteenth-century speculations on the eternal nature of the world played in seventeenth-century discourse (Bianchi, 1987). David Kubrin explicitly addresses the subject of Newton's cosmogony and has demonstrated that a cyclical view of time lies at the very heart of Newton's natural philosophy. In fact, Kubrin states that Newton's cosmogonical speculations were informed precisely by his rejection of the notion of an eternal world. Like many of his contemporaries, he shared the opposing belief in the progressive decline of the forces and regularity of the cosmos (Kubrin, 1967).

Newton's letter to Oldenburg on December 7, 1675 opens by reiterating his aversion to hypotheses and the "troublesome, insignificant disputes" that arise from them, and then continues by comparing the principles of electricity and magnetism to gravity. He suggests that the aetherial medium is composed of a "phlegmatic body" and "other various aetherial spirits." He goes so far as to declare that "perhaps the whole frame of nature may be nothing but [. . .] certain aetherial spirits, condensed as it were by precipitation" and that "thus perhaps may all things be originated from aether." On the basis of this hypothesis, the attraction of the Earth "may be caused by the continual condensation of some other such-like aetherial spirit, not of the main body of phlegmatic aether, but of something very thinly and subtly diffused through it, perhaps of an unctuous or gummy tenacious and springy nature." This spirit may penetrate and "be condensed in the pores of the Earth." The Earth's vast body "may continually condense so much of this spirit as to cause it from above to descend with great celerity for a supply" (Newton, 1953: 85–6; 1978: 252). Throughout the descent this spirit may bring with it the bodies it pervades with a force proportional to the surfaces of all the parts it acts upon. Nature, in fact, creates a circulation which, because of the slow ascent of a good deal of matter from the bowels of the Earth, "for a time constitutes the atmosphere, but being continually buoyed up by the new air, exhalations, and vapors rising underneath at length, (some part of the vapors which return in rain excepted) vanishes again into the aetherial spaces, and there perhaps in time relents and is attenuated into its first principle" (Newton, 1953: 86; 1978: 253).

The comparison of the Earth to a giant sponge that soaks up aether (the "active principle") which it gradually rids itself of is based on the assumption that nature "is a perpetual circulatory worker." Nature generates "fluids out of solids, and solids out of fluids; fixed things out of volatile, and volatile out of fixed; subtle out of gross, and gross out of subtle." Some substances rise from the center of the earth and "make the upper terrestrial juices, rivers, and the atmosphere" and, as a result, "others descend for a requital to the former."

What is true for the Earth may also be true for the sun. He, too, may "imbibe this spirit copiously, to conserve his shining and keep the planets from receding further from him." They who so desire it may also suppose that "the vast aetherial spaces between us and the stars are for a sufficient repository for this food of the sun and planets" (Newton, 1953: 86; 1978: 253).

In 1675 Newton was convinced that an "aether" was responsible for renewing motion and cosmic activity. In the *Principia* he suggested the comets were responsible: "For the conservation of the seas, and fluids of the planets, comets seem to be required, that, from their exhalations and vapors condensed, the wastes of their planetary fluids spent upon vegetation and putrefaction, and converted into dry earth, may be continually supplied and made up; And hence it is that the bulk of solid earth is continually increased and the fluids, if they are not supplied from without, must be in continual decrease and quite fail at last. I suspect, moreover, that it is chiefly from the comets that spirit comes, which is indeed the smallest but most subtle and useful part of our air, and so much required to sustain the life of all things with us" (Newton, 1965: 770–1; Kubrin, 1967: 336).

The need for active principles to perpetuate the universe required a mechanism through which the Creator could from time to time replenish the quantity of motion and the regularity of orbits. Newton believed the comets were such a mechanism. It not only explained the renewal of the amount of motion but also the continual, cyclical re-creation of the system and its subsequent development in time until the moment of the next creation (Kubrin, 1967: 345).

Chronology

Newton devoted a good deal of time and energy to a subject much discussed at the time; the reconciliation of Old Testament chronology to the secular world chronology of pagan or "gentile" peoples (Rossi, 1979). Newton had been studying the Christian religion and theology since the 1690s, and this eventually led to the painstakingly written *Chronology of Ancient Kingdoms Amended* (published in 1728, the year after his death), which included drafts and studies he had made many decades earlier. In keeping with late seventeenth- and early eighteenth-century religious tradition, the word "amended" meant the *abbreviation* of antiquity in order to avoid the heretical implication of Hermetic and libertine philosophers who believed in ancient kingdoms much older than the Hebrews as recounted in the Old Testament. According to these views, civilization, morality, and religion were not born of the exchange between God and Moses and God's delivery of the Ten Commandments to Moses. To acknowledge civilizations and populations pre-dating the Jews (Hermetics referring primarily to the Egyptians and libertines to Egyptian, Mexican, and Chinese cultures) would be to suggest that the Bible does not tell the story of the creation of the world and mankind but only that of a particular people; in turn suggesting that the Great Flood was not in fact a universal event but a local one that devastated a single population.

Newton (whose religious philosophy was, as we shall see, in many respects decisively heretical) did not disassociate himself from the many other defenders (Protestant and Catholic) of the veracity and singularity of the Old Testament. He insisted that the histories of all pagan societies and claims to a civilization older than the Israelites be compared to the Biblical narrative. Newton was one among many "abbreviators" of history. He wanted to show that the Jews predated the Greeks and other civilizations. He cut about 500 years from the accepted chronology of Greek history, got rid of a few thousand years from the timeline of other ancient populations, and primarily renewed and amplified an argument that would prove very popular: that the ancient civilizations revered by libertines who claimed they pre-dated the Jews had never existed and were but the product of what Giambattista Vico would refer to as "the arrogance of nations," or a society's attempt to make itself out as the oldest, and therefore the founder of all civilization. According to Newton, all nations had altered their histories in order to increase their status as the noblest of peoples. The gods of mythology as well as the kings and deified royalty of Chaldea, Assyria, and Greece have all been acknowledged as older than they really are. For this very reason, it was vanity that motivated the Egyptians to create the image of a kingdom thousands of years older than the world itself. All ancient civilizations (stated Newton along the lines of Bacon) are dubious, often invented, and always "full of poetical fictions": "The Egyptians boasted of a great and long-lived empire [. . .]. For purely vainglorious reasons, they made this kingdom out to be several thousands of years older than the world" (Newton, 1757: 144; 1779–85: V, 142–93).

Similar statements appear in *The Original of Monarchies* (1693–94 and published by Manuel): "Now all nations before they began to keep exact accounts of time have been prone to raise their antiquities and make the lives of their fathers longer than they really were [. . .]. For this made the Egyptians and the Chaldeans raise their antiquities higher than the truth by many thousands of years [. . .]. The Greeks and Latins are more modest in their own originals but yet have exceeded the truth" (Manuel 1963: 211). As Voltaire remarked in number seventeen of the *Lettres philosophiques*, all of the calculations Newton made based on the theory of the precession of the equinoxes and astronomical descriptions from ancient texts in order to establish dates served only to abbreviate the history of the world: "the distance between periods was shortened closer and everything happened later than we thought it had."

For those familiar with the work of Giambattista Vico, it is clear that Newton's "historical" writing reveals a great number of persistent and widespread ideas. It is truly a shame that so very few Vico scholars have ever read Newton and that just as few Newton scholars have more than simply glanced at Vico's *Scienza Nuova*.

Knowledge of the Ancients

In the 1963 book *Newton Historian*, Frank Manuel demonstrates how closely linked Newton's "physical history" of the universe was to "the history of nations." He points out that in Newton's world system a chronological event in the history of kingdoms could be translated into an astronomical event and vice versa because celestial and terrestrial events ran a parallel course. Just as "the formation of the planetary masses and the regulation of their movements had a temporal beginning, so the world was destined to end in a great conflagration as prophesied in the Book of the Apocalypse" (ibid.: 164).

Newton believed that pagan religious beliefs or Gentile theology had originated in Egypt. Gentile theology "was philosophical in nature and depended on astronomy and the physical science of the world system." He believed that Noah had settled in Egypt after the flood and that other lands had been settled when the sons fought over their inheritance and went their separate ways. Religion became identified with "the worship of a sacrificial fire that burned perpetually in the sanctuary of a sacred place." When Moses placed a sacred fire in the tabernacle he restored the original worship "purged of the superstitions introduced by the Egyptians." Such superstitions included making their ancestors into gods, and other societies all too willingly followed in their footsteps (Westfall, 1980: 353–4).

The anti-libertine polemic did not in fact exclude a belief in the myth of an ancient, primitive and received knowledge. Francis Bacon had presented his reform of learning as an *instauratio*, or the fulfillment of an ancient promise. The new science allowed man to regain the power he had once had over nature before the Fall. Bacon believed that "ancient legends" were neither products of their age nor the inventions of ancient poets but more like a "sacred relic and the gentle breezes from better days gone by, drawn from the oldest traditions of ancient civilizations and transferred to the pipes and trumpets of the Greeks" (Bacon, 1887–

92: VI, 627). That knowledge can be revived or resuscitated, that it is in some way hidden in the farthest reaches of time, and that basic truths existed long before Greek philosophy but were subsequently lost or concealed, are "Hermetic" ideas that ran through much of seventeenth-century culture and pop up again in the work of even the unlikeliest authors. Not only in Newton's work, as we shall see, but also for example in Descartes, who was a staunch believer in the superiority of modernity. In *Rules for the Direction of the Mind* he writes: "I am convinced that certain primary seeds of truth [. . .] thrived vigorously in that unsophisticated and innocent age of antiquity [. . .]. Some writers were able to grasp true ideas in philosophy and mathematics [. . .]. But I have come to think that these writers themselves, with a kind of pernicious cunning, later suppressed this mathematics as, notoriously, many inventors are known to have done where their own discoveries were concerned, fearing that their method, just because it was so easy and simple, would be deprecated if it were divulged (Descartes, 1897–1913: X, 376; 1985: 18).

In *De mundi systemate* (written between 1684 and 1686), Newton linked Copernican theory to not only Philolaus and Aristarchus but also Plato, Anaximander, and Numa Pompilius. In addition, he returned again to the idea of the ancient wisdom of the Egyptians: to symbolize the round orb with the sun at its center, Numa Pompilius erected a round temple to honor Vesta and decreed that a perpetual flame be kept in the middle of it. It is very likely, however, that this idea began with the Egyptians, the earliest astronomers. They in turn spread the soundest notions of philosophy abroad, in particular to the Greeks who tended more toward the study of philology than philosophy; "and we can even recognize in the worship of Vesta the spirit of the Egyptians who used concealed mysteries that were above the capacity of the common herd under the veil of religious rites and hieroglyphic symbols" (Newton, 1983a: 28–9; Westfall, 1980: 434–5).

The notion of a *prisca sapientia* is also found in the "Classical Scholia" which Newton had intended to add to the *Principia*. His aim was to prove that Greek and Italian philosophers in addition to Egyptian astronomers had known the phenomena and the laws of gravitational astronomy (ibid.). Newton, in fact, believed that it had always been known – if only symbolically – that the force of attraction diminished with respect to the square of distance: "The ancients did not adequately explain with what proportion gravity decreases as distance increases. Yet it seems they represented that proportion by celestial harmonies, using Apollo and his seven-stringed lyre to symbolize the sun and other six planets [. . .] and the intervals between the notes to represent the distance between the planets [. . .]. In the oracle of Apollo at Eusebius [. . .] the sun is called the King of seven-toned harmony. This symbol was used to show that the sun exerts its force on the planets [. . .] in inverse proportionate to the square of its distance from them" (ibid.: 143–4).

In all likelihood it is somewhat exaggerated to portray Newton as an "Hermetic" philosopher, yet he was without a doubt firmly convinced that he was *rediscovering* truths in natural philosophy that had been known since the dawn of time, revealed by God himself, concealed after the fall from grace, and then partially recovered in antiquity. The great Book of Nature had already been deciphered. Copernicus, Kepler, and Newton himself, had seen *progress* in astronomy also in terms of a *return* (McGuire and Rattansi, 1966).

Alchemy

Several thousands of handwritten pages composed over a lifetime are proof that Newton devoted a great deal of his efforts to the reading, transcription, and commentary of alchemical works. And if that were not enough, his papers show that he conducted numerous experiments using alkali, metals, and acids. When Newton linked gravity, as an active principle in the universe, to the cohesion of bodies and fermentation, we are reminded of his interests in chemistry and alchemy. From this perspective Newton's experiments surely served other purposes: to provide an experimental foundation for hypothetical inquiries into atoms and aether and for his attempt to find a unified explanation or unified science of the universe. This is evident in the last few lines of the *Principia*'s "General Scholium." In this section Newton refers to "a certain most subtle spirit which pervades and lies hidden in all gross bodies by the force and action of which spirit the particles of bodies attract one another at near distances and cohere, if contiguous; and electric bodies operate to greater distances [. . .]; and light is emitted [. . .] and all sensation is excited, and the members of animal bodies move at the command of the will by the vibrations of this spirit, mutually propagated [. . .] from the outward organs of the sense to the brain and from the brain to the muscles." Yet, concludes Newton, there is not "sufficiency of experiment which is required to an accurate determination and demonstration of the laws by which this electric and elastic spirit operates (Newton, 1953: 46–7; 1965: 796).

Newton had a longstanding interest in alchemy. In his late twenties he purchased some nitric acid, sublimate of mercury, antimony, alcohol, and saltpetre and singlehandedly built himself a brick furnace. At about the same time – 1669 – he began to read works by alchemists, attempting to penetrate their symbolic language and search for the rules and methods common to the various practitioners of the Art. Newton seems to have been more interested in the chemical aspects of alchemy than in the mystical–religious expressions that dominated the literature, and carried out many experiments during this time. Westfall reminds us that Newton approached the study of the Great Art with "unique intellectual equipment such as no other alchemist ever possessed." His interest in quantitative measurement was a predominant concern and he never abandoned the rigorous language born of mathematics in favor of metaphor alone. It is also true that Newton, from the beginning, had his reservations about the mechanical philosophy, believing its categories too confining to express the complexity of nature (Westfall, 1980: 293, 299, 301).

Westfall uses an ingenious metaphor to explain Newton's enduring interest in alchemy; an interest that disconcerted many scholars once his manuscripts on the subject became known. Westfall suggests that Newton's interest was a manifestation of *rebellion* against the confining limitations imposed by mechanical philosophy and he compares the relationship to a disenchanted married man who finds satisfaction in the company of another woman: "Mechanistic philosophy had surrendered to his desire, perhaps too readily. Unfulfilled, he continued the quest and found in alchemy, and in allied philosophies, a new mistress of infinite variety who never seemed fully to yield. Where others cloyed she only whet the appetite she fed. Newton wooed her in earnest for thirty years" (Westfall, 1980: 301).

Was alchemy just a long-term "extramarital affair" for Newton? This is difficult to accept when we connect his interest in alchemy to a series of other factors: his comments about how inopportune it was to openly present some of his theories; his convictions regarding "the end of the world"; his belief in an original, received knowledge that is a pure and uncorrupted truth; his comparison of the electric spirit, at times material and at others not, to a vital flame (Newton, 1991); and finally his remarks to Oldenburg about "aetherial spirits, condensed as it were by precipitation" and the constant "circular motion of Nature" (Newton, 1978: 252–3).

Newton's Religion and the Apocalypse

Newton believed in God and the Bible though he secretly held decidedly heretical views. He managed to carefully conceal his true ideas about Jesus Christ and Christianity all of his life, and in terms of his religious beliefs, he imitated Descartes by "advancing masked." It was quite remarkable that he managed to be exempted – by royal dispensation – from taking orders in the Anglican church since it was required of all fellows of Trinity College. In his later life, he spent many years editing possibly objectionable views from his theological writings with an eye to their eventual publication. On his deathbed Newton refused the sacrament of the Church in the presence of two witnesses (who never made it known) (Westfall, 1980: 330–4; 869).

Newton read many works written by the early Church fathers and was convinced (well before 1675) that Athanasius and his followers had perpetrated a massive fraud during the fierce conflict that characterized the history of the Church in the fourth century: the Holy Bible had been altered in a number of places. He believed the changes had been made to support the doctrine of Trinitarianism. In 1668 Newton was a fellow of the College of the *Holy and Undivided Trinity*, but he personally believed that Trinitarianism had been falsely imposed on Christians in the period of Athanasius's triumph over Arius. To worship Jesus Christ as God was, in his eyes, a manifestation of idolatry. The Roman Pope had sided with Athanasius and the Roman church, as a result, was the seat of an idolatrous sect begun *after* the early Church had established that only one true God was to be worshipped. Trinitarianism had become dogma for both the Roman Catholic and Anglican churches. By secretly professing himself an Arian, Newton recognized Christ as a divine mediator between God and man but not God: "The son confesseth the father is greater than him and calls him his God [. . .] the son in all things submits his will to the will of the father, which could be unreasonable if he were equal to the father." Christ should be worshipped as Lord, he added, "yet we are to do it without breaking the first commandment" (Westfall, 1980: 312, 313, 315–16, 824).

Christ was indeed the son of God but not God; he is not consubstantial to the Father. The two commandments at the heart of the religion – to love God and to love one's neighbor – "always have and always will be the duty of all nations and the coming of Jesus Christ has made no alteration in them." To love one's neighbor was taught to the pagans by Socrates, Cicero, and Confucius. The law of righteousness and charity "was dictated to the Christians by Christ, to the Jews by Moses and to all mankind by the light of reason" (Westfall, 1980: 821–2).

In many ways Newton's monotheistic Arianism was close to deism and libertine views of religion, and it was no coincidence that deism and Newtonianism appeared closely interrelated in the eighteenth century (Casini, 1980: 40). Newton wrote many more pages on theological themes than scientific ones. His commitment to such questions was so great that at several moments in his life he considered the problems of optics and physics an annoying interruption of more important work, namely the re-evaluation of all of Christianity (Westfall, 1980: 310).

Newton felt that it was important to carefully review Scripture and, in particular, the prophecies, in his study of the primitive Church. In fact, he was certain that he had reached a similar level of understanding the truths revealed by Biblical prophecy as he had the nature of colors and the laws of the universe: "Having searched (and by the grace of God obtained) after knowledge in the prophetic scriptures, I have thought myself bound to communicate it for the benefit of others, remembering the judgment of him who hid his talent in a napkin. [. . .] I would not have any discouraged by the difficulty and ill success that men have hitherto met with in these attempts. This is nothing but what ought to have been. For it was revealed to Daniel that the prophecies concerning the last times should be closed up and sealed until the time of the end: but then the wise should understand, and knowledge be increased. Dan 12: 4, 9, 10. And therefore the longer they have continued in obscurity, the more hopes there is that the time is at hand in which they are to be made manifest" (Newton, 1994: 3; Manuel, 1974: 107).

The reference to Daniel (Bacon also cites it on the title page of the *Novum Organum*) clearly shows that Newton believed he was living at the end of time when the meaning of the prophecies would inevitably be revealed. Even though late in life he revised his calculations of the second coming and moved it to the twentieth or twenty-first century, there can be no question that Newton proceeded from a millenarian perspective (Westfall, 1980: 816). The language of the prophecies, like that of nature, comes directly from God. Newton believed he had been elected by God and defined himself as one of the "few scattered people God had chosen, such as without being led by interest, education, or humane authorities, can set themselves sincerely and earnestly to search after truth" (Mamiani, 1990: 109; Westfall, 1980: 324–5).

Interpreting the Bible and Nature

Maurizio Mamiani has persuasively argued that even before formulating a somewhat coherent "scientific" theory, Newton had established a series of rules for interpreting Revelation. The *regulae philosophandi* in the *Principia* appear to be a sophisticated and simplified version of the rules Newton devised for interpreting the words and language of Scripture (Mamiani in Newton, 1994: xxix–xxxi). In the *Principia* Newton states that in constructing science we should not "recede from the analogy of Nature, which is wont to be simple and always consonant to itself." Newton had applied this same principle, years earlier, to his interpretation of the Holy Bible: "To observe diligently the content of Scriptures and analogy of the prophetic style [. . .]. To choose those constructions which without straining reduce things to the greatest simplicity. [. . .] Truth is ever to be found in simplicity, and not in multiplicity and confusion of things. As the world, which to the naked eye exhibits the greatest variety of objects, appears very simple in its internal constitution when surveyed by a philosophic understanding, and so much simpler by how much the better it is understood, so it is in these visions. It is the perfection of all God's works that they are all done with the greatest simplicity. He is the God of order and not of confusion" (Newton, 1994: 21, 29; Manuel, 1974: 116, 120).

The method used to interpret the Bible was essentially the same as that used to interpret nature. There was but one method for the discovery of truths and it applied to the Bible and natural philosophy alike; a property and feature of both science and religion. Like Galileo before him, Newton believed the two books – the Book of Scripture and the Book of Nature – could not contradict one another. Yet, in a departure from Galileo, he felt they must be read using the *very same rules*: As those who wish to understand the structure of the world must force themselves to reduce their knowledge to the simplest terms, so must it be in the search to understand the prophecies" (Newton, 1994: 29).

The Treatise on the Apocalypse begins with rules followed by definitions and propositions. The latter, as in the *Opticks*, "are tested in two ways through the rules and the definitions (equivalent to mathematical principles) and with direct reference to the Holy Bible (equivalent to an experiment or the comparison of phenomena)" (Mamiani, 1990: 110–11). Newton moreover believed that it was both possible and desirable to give a *scientific* reading of the Bible. In fact, a reading of the Bible according to the Newton's rules would be as reliable as a scientific fact: "Hence if any man shall contend that my Construction of the Apocalypse is uncertain, upon pretense that it may be possible to find out other ways, he is not to be regarded unless he shall show wherein what I have done may be mended. If the ways which he contends for be less natural or grounded upon weaker reasons, that very thing is demonstration enough that they are false, and that he seeks not truth but labors for the interest of a party." The analogy that immediately follows is even more striking: "For as of an Engine made by an excellent Artificer a man readily believes that the parts are right set together when he sees them joined truly with one [. . .] so a man ought with equal construction reason to acquiesce in that construction of the Prophecies when he sees their parts set in order according to their suitableness and the characters imprinted in them for that purpose. 'Tis true that an Artificer may make an Engine capable of being with equal congruity set together more ways than one, and that a sentence may be more ambiguous: but this Objection can have no place in the Apocalypse, because God who knew how to frame it without ambiguity intended it for a rule of faith" (Newton, 1994: 29–31; Manuel, 1974: 121).

Conclusion

As we have seen, a number of factors prevent us from coming to the hopelessly obsolete conclusion that Newton was a *positivist scientist* or even from praising him as the first great modern scientist. His lifelong interest in alchemy and firm belief in the existence of a primitive knowledge of origins together with his intertwining of science and religion, the concept of God and physics, and methods for the study of nature and the Bible place his work into an altogether different framework. Certainly modern science must have its heroes, and Newton is perhaps one of the greatest. The splendidly baroque inscription on his tomb is aptly expressive: "Let Mortals rejoice That there has existed such and so great an Ornament to the Human Race." The well-known couplet by Alexander Pope in some way also expresses a profound truth: "Nature and Nature's laws were hid in night, God said 'Let Newton be,' and all was light."

It is, however, also true that to construe all of Newton's statements as "modern" would be a hopeless task. This is not such an unpalatable conclusion to reach for one who, like myself, has spent the so-called "best years" of his life studying the relationship of magic to science during the period of the *birth* of modern science. Historians do not (and I believe should not) consider what we today call science a finished product but rather a series of attempts to contend with questions that were *at the time* unresolved and which, in many cases, were hard to accept as sensible and legitimate questions to pose.

The history of science can serve to make us aware of the fact that rationality, logical rigor, verifiable statements, the publicizing of results and methods, the very structure of scientific knowledge as something that can be built on are neither eternal categories of the spirit nor enduring facts in the history of mankind but historical achievements which, like all achievements, can by definition be lost or reversed.

As for the seemingly tumultuous origins of many of the values connected to scientific knowledge which we today assume as positive and incontestable, can we not identify a similar process relative to the political values of liberty and tolerance?

Chronology

YEAR	SCIENCE and TECHNOLOGY	POLITICS, RELIGION, and ART
1452–1519	Leonardo da Vinci, studies on mechanics and optics	
1482	Euclid is translated into Latin	
1492	Columbus discovers America	Bramante designs the apse of Santa Maria delle Grazie, Milan; Lorenzo de' Medici (the Magnificent) dies
1493–1541	Paracelsus (iatrochemistry)	
1494	Pacioli, *Summa arithmetica*	
1497		Leonardo da Vinci paints the *Last Supper*
1497–1500	Voyages of Vasco da Gama	
1498		Savonarola is burned at the stake in Florence
1501		Michelangelo sculpts the *David*
1509		Erasmus, *Praise of Folly*
1510–1511		The first African slaves are sent to America
1513–1521		Machiavelli, *The Prince; Discourses on the first decade of Livy*
1516		Ariosto, *Orlando Furioso*
1517		Luther posts his 95 theses
1519–1521	Magellan circumnavigates the world	Cortés conquers Mexico
1521		Luther is excommunicated
1527		The Sack of Rome

YEAR	SCIENCE and TECHNOLOGY	POLITICS, RELIGION, and ART
1529		Cambrai Peace Treaty between France and Spain
1530	Fracastoro, *Syphilis sive de morbo gallico*	Charles V is crowned Holy Roman Emperor
1533–1535		The city of Münster comes under Anabaptist control
1534		Society of Jesus is founded
1535		Sir Thomas More is beheaded
1536		Calvin, *Institutes of the Christian Religion*; Michelangelo paints the *Last Judgment*
1537	Latin translation of Apollonius of Perga	
1540	Biringuccio, *Pirotechnia*; Rheticus, *Narratio prima*	Calvin, the beginning of the Reformation in Geneva
1542	Fuchs, *De historia stirpium* (botanical treatise); Vesalius, *De fabrica corporis*	
1543	Copernicus, *De revolutionibus*	
1545	Cardano, *Ars Magna*	The Council of Trent is convoked
1546	Fracastoro, *De contagione*	
1551–1558	Gesner, *Historiae animalium*	
1551	Reinhold, *Tabulae prutenicae*	
1552	Cardano, *De subtilitate*	
1556	Agricola, *De re metallica*	
1558	Della Porta, *Magia naturalis*	
1562		Onset of the religious wars in France
1571		Battle of Lepanto
1572		St Bartholomew's day massacre of the Huguenots in Paris
1574	Tycho Brahe in Uraniborg	
1580	Palissy, *Discours admirables*	Montaigne, *Essays*
1582		Reform of the Gregorian calendar
1583	Cesalpino, *De plantis*	
1584	Bruno, *De l'infinito, universo e mondi*	Walter Raleigh establishes a colony on Roanoke Island, VA
1588	Brahe, *De mundi aetherie phaenomenis*	The defeat of the Spanish Armada

YEAR	SCIENCE and TECHNOLOGY	POLITICS, RELIGION, and ART
1589	Stevin's principles of mechanics	
1590, circa	Viète introduces the use of letters to algebra	
1590	Galileo, *De motu*	
1591		Shakespeare, *Henry VI*
1596	Kepler, *Mysterium cosmographicum*	
1598		The Edict of Nantes proclaims the religious and political freedom of the Huguenots
1599–1607	Aldovrandi's zoological encyclopedia is published	
1600	Gilbert, *De magnete*	Giordano Bruno is burned at the stake; Campanella, *The City of the Sun*; Shakespeare, *Hamlet*
1603	The Accademia dei Lincei is founded	
1605		Cervantes, *Don Quixote*
1609	Kepler, *Astronomia nova*	
1610	Galileo, *Sidereus nuncius*	The assassination of Henry IV, King of France
1611	Kepler, *Dioptrics*	
1615	Galileo, *Letter to Madama Cristina*	
1616	The Catholic Church condemns Copernicanism	
1618		The Thirty Years' War begins
1619	Kepler, *Harmonices mundi*	
1620	Bacon, *Novum organum*	
1623	Galileo, *Assayer*	
1625		Grotius, *De iure belli ac pacis*
1626	The "Jardin des plantes" is founded in Paris	
1627	Bacon, *New Atlantis*	
1629	Harvey, *De motu cordis*	
1632	Galileo, *Dialogue on the Two Chief Systems of the World*	Rembrandt, *The Anatomy Lesson*
1633	Galileo is tried by the Inquisition for heresy	
1635	Cavalieri states the principle of indivisibles	Calderón de la Barca, *La vida es sueño*

YEAR	SCIENCE and TECHNOLOGY	POLITICS, RELIGION, and ART
1637	Descartes, *Discourse on Method*	Corneille, *Le Cid*
1638	Fermat defines the method for finding the tangent to a curve; Galileo, *Dialogue on Two New Sciences*	
1640	Pascal, *Essai pour les coniques*	Comenius, *Didactica magna*; 25,000 colonists arrive in New England
1642		Hobbes, *De cive*; Civil War in England begins
1643	Torricelli's barometer experiment	
1647	Gassendi, *De vita Epicuri*; Pascal, *New Experiments*	Masaniello leads a popular revolt in Naples
1648	Van Helmont, *Ortus medicinae*	Peace of Westphalia brings an end to the Thirty Years War
1649	Descartes, *Passions of the Soul*	The execution of Charles I, King of England
1651	Guericke invents the pneumatic machine	Hobbes, *Leviathan*
1652–1654		England and Holland at war
1653–1658		The Protectorate of Oliver Cromwell
1656–1663		Bernini designs the colonnade of St Peter's Basilica
1657	The Accademia del Cimento is established; Huygens designs the pendulum clock	Naples plague
1661–1715		The reign of Louis XIV, King of France
1661	Boyle formulates the law of gases; Malpighi's microscope	
1662	The Royal Society is founded	
1665	Hooke, *Micrographia*; *Philosophical Transactions* begin publication	London plague
1666	Leibniz, *De arte combinatoria*; the Académie des Sciences is established in Paris and begins publishing the *Journal des Savants*; Magalotti's alcohol thermometer	Molière, *The Misanthrope*

YEAR	SCIENCE and TECHNOLOGY	POLITICS, RELIGION, and ART
1667		Milton, *Paradise Lost*
1668	Redi conducts experiments on spontaneous generation	
1669	Newton, *Methodus fluxionum*	
1670		Pascal, *Pensées*; Spinoza, *Treatise on Religious and Political Philosophy*
1671–1674	Malpighi studies the cellular structure of plant tissues	
1671	Kircher, *Ars magna*	
1672	Newton, *New theory of light and color*	
1675	Leméry, *Cours de chimie*	Vivaldi is born
1677	Leeuwenhoek studies spermatozoa under a microscope	Spinoza, *Ethics*
1679		England: the *Habeas Corpus Act* (safeguards against unlawful imprisonment)
1680	Borelli, *De motu animalium*	
1682	Ray, *Methodus plantarum nova*	
1683		Turks invade and hold Vienna
1684	Leibniz, *Nova methodus pro maximis et minimis*	
1685		Bach and Handel are born; Louis XIV revokes the Edict of Nantes
1687	Newton, *Principia*	
1688		The "Glorious Revolution" in England
1688–1713		Reign of Frederick I of Prussia
1689–1725		Reign of Peter the Great in Russia
1689		Locke, *Letters on Toleration*
1690		Locke, *Essay Concerning Human Understanding*
1694	Huygens, *Treatise on light*	
1695–1697		Bayle, *Dictionnaire*
1703		Leibniz, *New Essays on the Human Intellect*
1704	Newton, *Opticks*	

Bibliography

The first part of this bibliography lists only those works cited or specifically referred to in the chapters of this text. They are listed in alphabetical order by author's last name. The second part of the bibliography, "Additional reading," is a list by subject of important works not previously included.

Introduction

Bianchi, L. 1990. "L'esattezza impossibile: scienza e 'calculationes' nel XIV secolo, in L. Bianchi and E. Randi, *Le verità dissonanti* (Rome–Bari: Laterza), pp. 119–50.

——. 1997. *La filosofia nelle Università. Secoli XIII e XIV* (Firenze: La Nuova Italia).

Caspar, M. 1962. *Kepler: 1571–1630* (New York: Collier Books).

Clagett, M. 1959. *The Science of Mechanics in the Middle Ages* (Madison: University of Wisconsin Press).

De Libera, A. 1991. *Penser au Moyen Âge* (Paris: Éditions du Seuil).

Duhem, P. 1914–58. *Le système du monde*, 10 vols (Paris: Hermann).

Le Goff, J. 1977. "Quale coscienza l'Università medievale ha avuto di se stessa?" in *Tempo della Chiesa e tempo del mercante e altri saggi sul lavoro e la cultura del Medievo* (Turin: Einaudi), pp. 153–70.

——. 1993. *Intellectuals in the Middle Ages* (Cambridge, MA: Blackwell Publishers).

Mersenne, M. 1634. *Questions inouyes ou Récréation de Sçavants* (Paris).

Rossi, P. 1989. "Gli aristotelici e i moderni: le ipotesi e la natura," *La scienza e la filosofia dei moderni* (Turin: Bollati Boringhieri), pp. 90–113.

Westfall, R. S. 1980. *Never at Rest: A Biography of Isaac Newton* (Cambridge: Cambridge University Press).

White, L. Jr. 1962. *Medieval Technology and Social Change* (Oxford: Oxford University Press).

1 Obstacles

Agricola, G. 1563. *De l'arte de' metalli partita in dodici libri* (1556), translated into Tuscan by M. Michelangelo Florio Fiorentino (Basle: Hiernomino Frobenio and Nicolao Episcopo).

Bachelard, G. 1949. *Le rationalisme appliqué* (Paris: PUF).

——. 1938. *La formation de l'esprit scientifique* (Paris: Vrin).

Descartes, R. (Cartesio, R.). 1967. *Opere*, ed. E. Garin, 2 vols. (Bari: Laterza).

Guidobaldo del Monte 1581. *Le Mechaniche*, translated by F. Pigafetta (Venice).

Koyré, A. 1971. *Mystiques, spirituels, alchimistes du XVIe siècle* (Paris: Gallimard).

Kuhn, T. 1980. "The Halt and the Blind: Philosophy and the History of Science," in *The British Journal for the Philosphy of Science*, XXXI, pp. 181–92.

2 Secrets

Agricola, G. 1563. *De l'arte de' metalli partita in dodici libri* (1556), translated into Tuscan by M. Michelangelo Florio Fiorentino (Basle: Hiernomino Frobenio and Nicolao Episcopo).

Agrippa, C. 1550. *Opera omnia*, 2 vols. (Lyon).

Bacon, F. 1887–92. *Works*, ed. R. L. Ellis, J. Spedding, and D. D. Heath, 7 vols. (London).

Biringuccio, V. 1558. *De la Pirotechnia libri dieci* (1540) (Venice).

Bono Da Ferrara, P. 1602. *Introductio in artem chemiae* (Montisbeligardi).

Comenius, G. A. 1974. *Opere*, ed. M. Fattori (Turin: Utet).

Eamon, W. 1990. "From the Secrets of Nature to Public Knowledge," in D. C. Lindberg and R. S. Westman, *Reappraisals of the Scientific Revolution* (Cambridge: Cambridge University Press), pp. 333–65.

Fracastoro, G. 1574. *Opera omnia* (Venice).

Gilbert, W. 1958. *De Magnete*, ed. P. F. Mottelay (New York: Dover).

Maldonado, T. 1991. "Il brevetto tra invenzione e innovazione," in *Rassegna*, XIII, pp. 6–11.

Mersenne, M. 1625. *La vérité des sciences* (Paris).

Perrone Compagni, V. 1975. "Picatrix Latinus: concezioni religiose e prassi magica," in *Medioevo*, I, pp. 237–337.

Sprat, T. 1667. *The History of the Royal Society of London for the Improving of Natural Knowledge* (London).

Taylor, F. S. 1949. *A Survey of Greek Alchemy* (New York).

Thorndike, L. 1923. *The History of Magic and Experimental Science*, 8 vols. (New York: Columbia University Press).

Vaughan, T. 1888. *The Magical Writings of Th. Vaughan*, ed. A. E. Waite (London).

3 Engineers

Agricola, G. 1563. *De l'arte de' metalli partita in dodici libri* (1556), translated into Tuscan by M. Michelangelo Florio Fiorentino (Basle: Hiernomino Frobenio and Nicolao Episcopo).
Antal, F. 1947. *Florentine Painting and its Social Background* (London: Kegan Paul).
Bacon, F. 1975. *Scritti filosofici*, ed. P. Rossi (Turin: Utet).
Barbaro, D. 1556. *I dieci libri dell'architettura di Vitruvio tradotti e commentati* (Venice).
Biringuccio, V. 1558. *De la Pirotechnia libri dieci* (1540) (Venice).
Brizio, A. M. 1954. *Leonardo, saggi e ricerche* (Rome).
Diderot, D. 1875–7. *Oeuvres complètes*, 20 vols. (Paris).
Dürer, A. 1528. *Vier Bücher von menschlicher Proportion* (Nuremberg).
Lorini, B. 1597. *Delle fortificazioni* (Venice).
Luporini, C. 1953. *La mente di Leonardo da Vinci* (Florence: Sansoni).
Norman, R. 1581. *The New Attractive* (London).
Palissy, B. 1880. *Oeuvres* (Paris).
Paré, A. 1840–1. *Oeuvres*, 3 vols. (Paris).
Ramelli, A. 1588. *Le diverse et artificiose macchine* (Paris).
Rossi, P. 1970. *Philosophy, Technology, and the Arts in the Early Modern Era* (New York: Harper and Row).
——. 1971. *I filosofi e le macchine: 1400–1700* (Milan: Feltrinelli).
Solmi, E. (ed.). 1889. *Frammenti letterari e filosofici di Leonardo da Vinci* (Florence).
Vitruvius. 1556. *I dieci libri dell'architettura di Vitruvio tradotti e commentati da Monsignore Barbaro* (Venice: for Francesco Marcolini).

4 The Unseen World

Bacon, F. 1975. *Scritti filosofici*, ed. P. Rossi (Turin: Utet).
——. 1887–92. *Works*, ed. E. R. Ellis, J. Spedding, and D. D. Heath, 7 vols. (London).
Borel, P. 1657. *Discours nouveaux prouvant la pluralité des mondes* (Geneva).
Campanella, T. 1941. *La Città del Sole*, ed. N. Bobbio (Turin: Einaudi).
Febvre, L. and Martin, H. J. 1958. *L'Apparition du livre* (Paris: Albin Michel).
Fuchs, L. 1542. *De historia stirpium* (Basle).
Gombrich, E. H. 1950. *The Story of Art* (London: Phaidon Press).
——. 1960. *Art and Illusion* (New York: Pantheon Books).
Hall, A. R. 1954. *The Scientific Revolution, 1500–1800* (London: Longmans, Green and Co.).
Hooke, R. 1665. *Micrographia* (London).
McLuhan, M. 1964. *Understanding Media* (New York: McGraw-Hill Book Co.).
Montaigne, M. de 1970. *Saggi*, ed. F. Garavini (Milan: Mondadori).
Panofski, E. 1943. *Albrecht Dürer* (Princeton: Princeton University Press).

Pascal, B. 1959. *Opuscoli e scritti vari*, ed. G. Preti (Bari: Laterza).

Steinberg, S. H. 1955. *Five Hundred Years of Printing* (Harmondsworth: Penguin Books).

Thorndike, L. 1957. "Newness and Novelty in Seventeenth-Century Science," in P. Weiner and A. Noland, *Roots of Scientific Thought* (New York: Basic Books).

Vesalius, A. 1964. Preface to *Fabbrica* and Letter to G. Oporini, ed. L. Premuda (Padua: Liviana).

Wiener, P. and Noland, A. 1957. *Roots of Scientific Thought* (New York: Basic Books).

5   A New Universe

Brahe, T. 1913–29. *Opera omnia*, ed. J. L. E. Dreyer, 15 vols. (Copenhagen: Libraria Gyldendaliana).

Camporeale, S. 1977–8. "Umanesimo e teologia tra '400 e '500," in *Memorie Domenicane*, New Series.

Copernicus, N. 1979. *Opere*, ed. F. Barone (Turin: Utet).

Donne, J. 1933. *Poems*, ed. H. J. C. Grierson (London).

Galilei, G. 1953. *Dialogue Concerning the Two Chief World Systems: Ptolemaic and Copernican*, ed. S. Drake, (Berkeley, CA: University of California Press).

Garin, E. 1975. *Rinascite e rivoluzioni. Movimenti culturali dal XIV al XVIII secolo* (Rome–Bari: Laterza).

Kepler, J. 1858–71. *Opera omnia*, ed. C. Frisch, 8 vols. (Frankfurt: Heyder und Zimmer).

——. 1937. *Gesammelte Werke*, ed. M. Caspar (Munich: Beck).

Koyré, A. 1961. *La révolution astronomique, Copernic, Kepler, Borelli* (Paris: Hermann).

Kuhn, T. 1957. *The Copernican Revolution* (Cambridge, MA: Harvard University Press).

Rheticus, G. 1541. *De libris revolutionum N. Copernici narratio primo* (Basle).

Westfall, R. S. 1971. *The Construction of Modern Science. Mechanisms and Mechanics* (New York: John Wiley and Sons).

6   Galileo

Acta 1979. "Acta Ioannis Pauli PP," in *Acta Apostolicae Sedis*, LXXI.

Clavelin, M. 1968. *La philosophie naturelle de Galilée* (Paris: Colin).

Galilei, G. 1890–1909. *Opere*, 20 vols. (Florence: Barbera).

——. 1953. *Dialogue Concerning the Two Chief World Systems: Ptolemaic and Copernican*, ed. S. Drake (Berkeley, CA: University of California Press).

——. 1957. *Discoveries and Opinions of Galileo*, ed. S. Drake (New York: Doubleday Anchor Books).

Poppi, A. 1992. *Cremonini e Galilei inquisiti a Padova nel 1604* (Padua: Antenore).

Schmitt, C. 1969. "Experience and Experiment: A Comparison of Zabarella's

View with Galileo's in *De Motu*," in *Studies in the Renaissance*, XVI, pp. 80–138.

Shea, W. R. 1972. *Galileo's Intellectual Revolution* (London: Macmillan).

Westfall, R. S. 1971. *The Construction of Modern Science: Mechanisms and Mechanics* (New York: John Wiley and Sons).

Wisan, W. 1974. "The New Science of Motion: A Study of Galileo's *De Motu Locali*," in *Archive for the History of Exact Sciences*, ed. C. Truesdell, XIII, nos 2–3.

7 Descartes

Descartes, R. 1897–1913. *Oeuvres*, eds C. Adam and P. Tannery, 12 vols and supplement (Paris: Cerf).

——. 1966–83. *Opere Scientifiche*, vol. I, ed. G. Micheli (Turin: Utet); vol. II, ed. E. Lojacono (Turin: Utet).

——. 1967. *Opere*, ed. E. Garin, 2 vols (Bari: Laterza).

——. 1985. *The Philosophical Writings of Descartes*, eds J. Cottingham, R. Stoothoff and D. Murdoch, 3 vols (Cambridge: Cambridge University Press).

Koyré, A. 1965. *Newtonian Studies* (Cambridge, MA: Harvard University Press).

Shea, W. 1991. *The Magic of Number and Motion: The Scientific Career of René Descartes* (Nantucket, MA: Watson Publishing International).

Westfall, R. S. 1971. *The Construction of Modern Science: Mechanisms and Mechanics* (New York: John Wiley and Sons).

8 Countless Other Worlds

Borel, P. 1657. *Discours nouveau prouvant la pluralité des mondes* (Geneva).

Bruno, G. 1907. *Opere italiane*, vol. I, "Dialoghi metafisici," ed. G. Gentile (Bari: Laterza).

Campanella, T. 1994. *A Defense of Galileo*, translated with an introduction and notes by R. J. Blackwell (Notre Dame, IN: University of Notre Dame Press).

Cusa, N. 1932. *Opera omnia*, ed. E. Hoffmann and R. Klibanski (Leipzig).

Descartes, R. 1936–63. *Correspondences*, ed. C. Adam and J. Milhaud, 8 vols. (Paris: PUF).

——. 1967. *Oeuvres*, eds C. Adam and P. Tannery, 12 vols and supplement (Paris: Cerf).

Galilei, G. 1890–1909. *Opere*, 20 vols (Florence: Barbera).

Huygens, C. 1698. *Cosmotheoros sive de terris coelestibus earumque ornatu conjecturae* (Hagae Comitorum).

——. 1888–50. *Oeuvres complètes*, 22 vols (The Hague: Martinus Nijhoff).

Kepler, J. 1858–71. *Opera omnia*, ed. C. Frisch (Frankfurt: Heyder und Zimmer).

——. 1937–59. *Gesammelte Werke*, ed. M. Caspar, 18 vols (Munich: Beck).

——. 1967. *Somnium*, ed. E. Rosen (Madison: University of Wisconsin).

——. 1972. *Dissertatio e Narratio*, ed. E. Pasoli and G. Tabarroni (Turin: Bottega d'Erasmo).

Koyré, A. 1957. *From the Closed World to the Infinite Universe* (Baltimore, MD: Johns Hopkins University Press).

Kuhn, T. 1957. *The Copernican Revolution* (Cambridge, MA: Harvard University Press).

Lovejoy, A. O. 1936. *The Great Chain of Being* (Cambridge, MA: Harvard University Press).

Nicolson, M. 1960. *Voyages to the Moon* (New York: Macmillan).

Wilkins, J. 1638. *Discovery of a New World and Another Planet* (London: Rich Baldwin).

9 Mechanical Philosophy

Bacon, F. 1887–92. *Works*, ed. E. R. Ellis, J. Spedding, and D. D. Heath, 7 vols (London).

——. 1975. *Scritti filosofici*, ed. P. Rossi (Turin: Utet).

Borelli, G. A. 1680–81. *De motu animalium, opus posthumum pars prima et secunda* (Rome).

Boyle, R. 1772. *The Works of the Honorable R. Boyle*, ed. T. Birch, 7 vols (London).

Descartes, R. 1897–1913. *Oeuvres*, ed. C. Adam and P. Tannery, 12 vols and supplement (Paris: Cerf).

——. 1967. *Opere*, ed. E. Garin, 2 vols (Bari: Laterza).

——. 1985. *The Philosophical Writings of Descartes*, eds J. Cottingham, R. Stoothoff, and D. Murdoch, 3 vols (Cambridge: Cambridge University Press).

Dijksterhuis, E. J. 1961. *The Mechanization of the World Picture* (Oxford: Clarendon Press).

Gassendi, P. 1649. *Syntagma philosophiae Epicuri cum refutationibus dogmatum quae contra fidem christianam ab eo asserta sunt* (Lyon).

——. 1658–75. *Opera omnia*, 6 vols (Lugduni).

Hobbes, T. 1839–45. *Opera philosophica*, ed. W. Molesworth, 5 vols (London).

——. 1950. *Leviathan* (New York: Dutton).

Hooke, R. 1665. *Micrographia* (London).

——. 1705. *Posthumous Works* (London).

Laudan, H. 1981. *Science and Hypothesis: Historical Essays on Scientific Methodology* (Dordrecht: Reidel).

Leibniz, G. W. 1840. *Opera philosophica*, ed. I. Erdmann, 2 vols (Berlin).

——. 1849–63. *Mathematische Schriften*, ed. C. I. Gerhardt, 7 vols (Berlin–Halle).

——. 1875–90. *De philosophischen Schriften*, ed. C. I. Gerhardt, 7 vols (Berlin).

Malpighi, M. 1944. *De pulmonibus seguito dalla Risposta apologetica*, ed. S. Baglioni (Rome).

Mersenne, M. 1636. *Harmonie Universelle* (Paris).

Newton, I. 1721. *Opticks* (London).

——. 1953. *Newton's Philosophy of Nature: Selections from His Writings*, ed. H. S. Thayer (New York: Hafner Press).

——. 1962. *Unpublished Scientific Papers. A Selection from the Portsmouth*

Collection, ed. A. R. Hall and M. Boas Hall (Cambridge MA: Cambridge, University Press).

Vico, G. B. 1957. *Tutte le opere* (Milan: Mondadori).

Westfall, R. S. 1971a. *Force in Newton's Physics: The Science of Dynamics in the Seventeenth Century* (London: Macdonald and Co.).

——. 1971b. *The Construction of Modern Science: Mechanisms and Mechanics* (New York: John Wiley and Sons).

10 Chemical Philosophy

Abbri, F. 1978. *La chimica nel Settecento* (Turin: Loescher).

——. (ed.) 1980. *Elementi, principi, particelle: le teorie chimiche da Paracelso a Stahl* (Turin: Loescher).

——. 1984. *Le terre, le acque, le arie: la rivoluzione chimica del Settecento* (Bologna: Il Mulino).

Beguin, J. 1665. *Les éléments de chymie* (Lyon: Claude La Rivière).

Boyle, R. 1900. *The Sceptical Chemist* (New York: E. P. Dutton & Co.).

——. 1772. *The Works*, ed. Th. Birch, 7 vols (London).

Debus, A. G. 1970. *Science and Education in the Seventeenth Century* (London).

——. 1977. *The Chemical Philosophy* (New York: Watson).

Leméry, N. 1682. *Cours de chimie* (1675) (Paris).

——. 1922–33. *Samtliche Werke*, ed. K. Sudhoff (Munich).

Paracelsus. 1973. *Paragrano*, ed. F. Masini (Rome–Bari: Laterza).

Partington, J. R. 1961–2. A *History of Chemistry*, vols 2 and 3 (London: Macmillan).

Quercetanus 1684. *Le ricchezze della riformata farmacopea. Nuovamente di favella latina trasportata in italiano dal sig. G. Ferrari* (Venice).

Stahl, G. 1783. *Traité des Sels dans lequel on démontre qu'ils sont composés d'une Terre subtile, intimement combiné avec de l'eau. Traduit de l'Allemand* (Paris).

Webster, C. 1982. *From Paracelsus to Newton: Magic and the Making of Modern Science* (Cambridge: Cambridge University Press).

Westfall, R. S. 1971. *The Construction of Modern Science: Mechanism and Mechanics* (New York: John Wiley and Sons).

11 Magnetic Philosophy

Cabeo, N. 1629. *Philosophia magnetica* (Ferrara).

De Martino, E. 1961. *La terra del rimorso: contributo ad una storia religiosa del Sud* (Milan: Il Saggiatore).

Descartes, R. 1936–63. *Correspondence*, ed. C. Adam and G. Milhaud, 8 vols (Paris: PUF).

Dijksterhuis, E. J. 1961. *The Mechanization of the World Picture* (Oxford: Oxford University Press).

Galilei, G. 1890–1909. *Opere*, 20 vols (Florence: Barbera).
Gilbert, W. 1958. *De Magnete*, ed. P. F. Mottelay (New York: Dover).
Heilbron, J. L. 1979. *Electricity in the 17th and 18th Centuries* (Berkeley, CA: University of California Press).
——. 1982. *Elements of Early Modern Physics* (Berkeley, CA: University of California Press).
Kircher, A. 1654. *Magnes sive de arte magnetica opus tripartitum* (Rome).
Magalotti, L. 1806. *Saggi di naturali esperienze fatte nell'Accademia del Cimento* (Milan).
——. 1976. *Saggi di naturali esperienze*, ed. T. Poggi Salani (Milan: Longanesi).
Muraro, L. 1979. *Giambattista della Porta mago e scienziato* (Milan: Feltrinelli).
Nocenti, L. 1991. "Athanasius Kircher, mago ed enciclopedista" (thesis, Department of Philosophy, University of Florence, academic year 1990–91).
Shea, W. 1991. *The Magic of Numbers and Motion: The Scientific Career of René Descartes* (Nantucket, MA: Watson Publishing International).

12   The Heart and Generation

Adelmann, H. D. 1966. *Marcello Malpighi and the Evolution of Embryology* (Ithaca, NY: Cornell University Press).
Bernardi, W. 1980. *Filosofia e scienze della vita: la generazione animale da Cartesio a Spallanzani* (Turin: Loescher).
——. 1986. *Le metafisiche dell'embrione: scienze della vita e filosofia da Malpighi a Spallanzani* (Florence: Olschki).
Dobell, C. 1932. *Anthony van Leeuwenhoek and His Little Animals* (London: I. Bale and Danielsson).
Pagel, W. 1967. *William Harvey's Biological Ideas* (New York–Basle: S. Karger).
Redi, F. 1668. *Esperienze intorno alla generazione degli insetti* (Florence: for Piero Matini).
Roger, J. 1963. *Les Sciences de la vie dans la pensée française du XVIIIe siècle. La génération des animaux de Descartes à l'Encyclopédie* (Paris: Colin).
Solinas, G. 1967. *Il microscopio e le metafisiche: epigenesi e preesistenza da Cartesio a Kant* (Milan: Feltrinelli).

13   Time in nature

Burnet, T. 1684. *The Sacred History of the Earth* (London).
Hooke, R. 1705. "Discourse on Earthquakes" (1668), in *The Posthumous Works* (London: Waller).
Leibniz, G. W. 1749. *Protogaea, sive de prima facie Telluris dissertatio* (Goettingen).
Rossi, P. 1979. *I segni del tempo: storia della Terra e storia delle nazioni da Hooke a Vico* (Milan: Feltrinelli).
Rudwick, M. J. S. 1976. *The Meaning of Fossils: Episodes in the History of Paleontology* (New York: Watson).

Scilla, A. 1670. *La vana speculatione disingannata dal senso. Lettera responsiva circa i corpi marini che pietrificati si truovano in vari luoghi terrestri* (Naples: Andrea Colicchia).

Solinas, G. 1973. *La "Protogaea" di Leibniz, ai margini della rivoluzione scientifica* (Istituto di Filosofia della Facoltà di Lettere e Filosofia, University of Cagliari).

Steensen, N. (Nicolaus Steno or Stenone) 1669. *De solido intra solidum naturaliter contento dissertationis prodromus* (Florence).

14 Classification

Eco, U. 1995. *The Search for the Perfect Language* (Oxford: Blackwell Publishers).

Emery, C. 1948. "John Wilkin's Universal Language," *Isis*, pp. 174–85.

Fontenelle, B. de 1708. "Eloge de M. de Tournefort," in *Histoire de l'Académie des Sciences* (Paris).

Linnaeus, C. von 1766. "Necessitas historiae naturalis Rossiae, quam, Preside C. Linnaeo, proposuit Alexander de Karamyschew," Uppsala, May 15, in C. von Linnaeus, *Amoenitates Academicae*, VII (Stockholm), 1769.

——. 1784. *Philosophica botanica* (Cologne).

Luria S. E., Gould S. J., and Singer, S. 1981. *A View of Life* (Menlo Park, CA: Benjamin Cummings Publishing).

Ray, J. 1718. *Philosophical Letters* (London).

——. 1740. *Selected Remains of the Learned John Ray by the Late William Derham* (London).

Rossi, P. 1983. *Clavis Universalis: arti della memoria e logica combinatoria da Lullo a Leibniz* (Bologna: Il Mulino).

——. 1984. "Universal Languages, Classification, and Nomenclatures in the Seventeenth Century," *History and Philosophy of Life Sciences*, II, pp. 119–31.

Tournefort, J. Pitton de 1700. *Institutiones rei herbariae*, 3 vols (Paris).

——. 1797. *Éléments de botanique ou méthode pour connaître les plantes*, 6 vols (Lyon).

Wilkins, J. 1668. *An Essay Towards a Real Character and a Philosophical Language* (London).

Yates, F. 1966. *The Art of Memory* (London: Routledge and Kegan Paul).

15 Instruments and Theories

Berkeley, G. 1996. *Opere filosofiche*, ed. S. Parigi (Turin: Utet).

Castelnuovo, G. 1962. *Le origini del calcolo infinitesimale nell'era moderna* (Milan: Feltrinelli).

Dijksterhuis, E. J. 1961. *The Mechanization of the World Picture* (Oxford: Oxford University Press).

Feynman, R. T., Leighton, R. B., and Saads, M. 1963. *The Feynman Lectures on Physics* (Reading, MA: Addison-Wesley).

Gallino, L. 1995. "La costruzione della conoscenza scientifica in Netropolis," paper given at the Conference "Ricerca scientifica e communicazione nell'età della telematica," Centro Studi Telecom Italia, Venice, March 10.

Giorello, G. 1985. *Lo spettro e il libertino: teologia, matematica, libero pensiero* (Milan: Mondadori).

Giusti, E. 1984. "A tre secoli dal calcolo: la questione delle origini," *Bollettino dell'Unione Matematica Italiana*, VI, 3A, pp. 1–55.

——. 1989. "Il calcolo infinitesimale tra Leibniz e Newton," in G. Tarozzi and M. van Vloten, eds, *Radici, significato, retaggio dell'opera newtoniana* (Bologna: Società Italiana di Fisica), pp. 279–95.

——. 1990. Introduction to G. Galilei, *Discorsi e dimostrazioni matematiche intorno a due nuove scienze* (Turin: Einaudi).

Grant, E. 1981. *Much Ado about Nothing: Theories of Space and Vacuum from the Middle Ages to the Scientific Revolution* (Cambridge: Cambridge University Press).

Hacking, I. 1983. *Representing and Intervening* (Cambridge: Cambridge University Press).

Hall, A. R. 1980. *Philosophers at War: The Quarrel between Newton and Leibniz* (Cambridge: Cambridge University Press).

Kline, M. 1953. *Mathematics in Western Culture* (New York: Oxford University Press).

Lombardo Radice, L. 1981. *L'infinito* (Rome: Editori Riuniti).

Pickering, A. (ed.) 1992. *Science as Practice and Culture* (Chicago: University of Chicago Press).

Shapin, S. and Schaffer, S. 1985. *Leviathan and the Air-Pump: Hobbes, Boyle, and the Experimental Life* (Princeton, NJ: Princeton University Press).

Singh, J. 1959. *Great Ideas of Modern Mathematics* (New York: Dover).

Westfall, R. S. 1980. *Never at Rest: A Biography of Isaac Newton* (Cambridge: Cambridge University Press).

Wieland, W. 1970. *Die aristotelische Physik* (Goettingen: Vandenhoek & Ruprecht).

16   Academies

n.a. 1981. "Accademie scientifiche del Seicento. Professioni borghesi," in *Quaderni Storici*, XVI, 48.

Altieri Biagi, M. L. 1969. *Scienziati italiani del Seicento* (Milan–Naples: Ricciardi).

Arnaldi, G. (ed.). 1974. *Le origini dell'Università* (Bologna: Il Mulino).

Ben David, J. 1971. The Scientist's Role in Society (Englewood Cliffs, NJ: Prentice-Hall).

Bertrand, J. 1869. *L'Académie Royale des Sciences de 1666 à 1793* (Paris).

Boehm, L. and Raimondi, E. (eds). 1981. *Università, accademie e società scientifiche in Italia e in Germania dal Cinquecento al Settecento* (Bologna: Il Mulino).

Borselli, L., Poli, C., and Rossi, P. 1983. "Una libera comunità di dilettanti nella

Parigi del Seicento," in *Cultura dotta e cultura popolare nel Seicento* (Milan: Franco Angeli).

Crosland, M. P. (ed.). 1975. *The Emergence of Science in Western Europe* (London-Basingstoke: Macmillan).

Dallari, U. 1888–1924. *I rotuli dei lettori leggisti e artisti dello studio bolognese dal 1384 al 1799* (Bologna).

Farrar, W. V. 1975. "Science and the German University System,"in Crosland, 1975, pp. 179–92.

Galluzzi, P. 1981. "L'Accademia del Cimento," *Accademie scientifiche*, pp. 788–844.

Hackmann, W. D. 1975. "The Growth of Science in the Netherlands in the Seventeenth and Early Eighteenth Centuries," in Crosland, pp. 89–110.

Hahn, R. 1971. *The Anatomy of a Scientific Institution: The Paris Academy of Sciences (1666–1803)* (Berkeley, CA: University of California Press).

Hall, A. R. 1954. *The Scientific Revolution: 1500–1800* (London: Longmans).

——. 1963. *From Galileo to Newton: 1630–1720* (London: Collins).

Hammerstein, N. 1981. "Accademie e società scientifiche in Leibniz," in Boehm and Raimondi, pp. 395–419.

Johnson, F. J. 1957. "Gresham College: Precursor of the Royal Society," in Weiner and Noland, pp. 328–53.

Kraft, F. 1981. "Luoghi della ricerca naturale," in Boehm and Raimondi, pp. 421–60.

Mathias, P. (ed.). 1972. *Science and Society: 1600–1900* (Cambridge: Cambridge University Press).

Olmi, G. 1981. "Federico Cesi e i Lincei," in Boehm and Raimondi, pp. 169–237.

Quondam, A. 1981. "La scienza e l'accademia," in Boehm and Raimondi, pp. 21–68.

Schmitt, B. 1975. "Science in the Italian Universities in the Sixteenth and Early Seventeenth Centuries," in Crosland, pp. 35–56.

Sprat, T. 1966. *A History of the Royal Society of London* (1667) (London: Kegan Paul).

Tega, W. (ed.). 1986. *Anatomie accademiche: I, I Commentari dell'Accademia delle Scienze di Bologna* (Bologna: Il Mulino).

Torrini, M. 1981. "L'Accademia degli Investiganti," *Accademie scientifiche*, pp. 845–83.

Weiner, P. and Noland, A. (eds). 1957. *Roots of Scientific Thought* (New York: Basic Books).

Westfall, R. S. 1971. *The Construction of Modern Science: Mechanism and Mechanics* (New York: John Wiley and Sons).

17   Newton

Bacon, F. 1887–92. *Works*, ed. R. L. Ellis, J. Spedding, and D. D. Heath, 7 vols (London).

Bevilacqua, F. and Ianniello, M. G. (eds). 1982. *L'ottica dalle origini all'inizio del Settecento* (Turin: Loescher).

Bianchi, L. 1987. *L'inizio dei tempi: antichità e novità del mondo da Bonaventura a Newton* (Florence: Olschki).

Casini, P. 1980. *Introduzione all'Illuminismo* (Bari–Rome: Laterza).

Descartes. R. 1897–1913. *Oeuvres*, ed. C. Adam and P. Tannery, 12 vols (Paris: Cerf).

——. 1985. *The Philosophical Writings of Descartes*, ed. J. Cottingham, R. Stoothof, and D. Mundoch, 3 vols (Cambridge: Cambridge University Press).

Feynman, R. T., Leighton, R. B., and Saads, M. 1963. *The Feynman Lectures on Physics* (Reading, MA: Addison-Wesley).

Hall, A. R. 1980. *Philosophers at War: The Quarrel between Newton and Leibniz* (Cambridge: Cambridge University Press).

Herivel, J. 1965. *The Background to Newton's Principia* (Oxford: Oxford University Press).

Koyré, A. 1965. *Newtonian Studies* (Cambridge, MA: Harvard University Press).

Kubrin, D. 1967. "Newton and the Cyclical Cosmos: Providence and the Mechanical Philosophy," *Journal of the History of Ideas*, vol. 8, no. 3 (July–September), pp. 325–46.

Leibniz–Clarke 1956. *The Leibniz–Clarke Correspondence*, ed. H. Alexander (New York).

Mamiani, M. 1990. *Introduzione a Newton* (Rome–Bari: Laterza).

Manuel, F. 1963. *Newton Historian* (Cambridge: Cambridge University Press).

——. 1974. *The Religion of Isaac Newton* (Oxford: Clarendon Press).

McGuire, J. E. and Rattansi, P. M. 1966. "Newton and the 'Pipes of Pan'," *Notes and Records of the Royal Society of London*, XXI, pp. 108–43.

Newton, I. 1721. *Opticks* (London).

——. 1757. *La cronologia degli antichi regni emendata*, Italian translation by P. Rolli (Venice).

——. 1779–85. *Opera quae extant omnia*, 5 vols (London).

——. 1953. *Newton's Philosophy of Nature: Selections from his Writings*, ed. H. S. Thayer (New York: Hafner Press).

——. 1959–77. *The Correspondence of Isaac Newton*, ed. H. W. Turnbull, J. P. Scott, A. R. Hall, and L. Tilling, 7 vols (Cambridge).

——. 1962. *Unpublished Scientific Papers of I. Newton*, ed. A. R. Hall and M. Boas Hall (Cambridge: Cambridge University Press).

——. 1965. *Principi matematici della filosofia naturale*, ed. A. Pala (Turin: Utet).

——. 1967–81. *The Mathematical Papers of I. Newton*, ed. D. T. Whiteside, 8 vols (Cambridge: Cambridge University Press).

——. 1978. *Scritti di ottica*, ed. A. Pala (Turin: Utet).

——. 1983a. *Il sistema del mondo e gli scolii classici*, ed. P. Cassini (Rome: Theoria).

——. 1983b. *Certain Philosophical Questions: Newton's Trinity Notebook*, ed. J. E. McGuire and M. Tamny (Cambridge: Cambridge University Press).

——. 1984. *The Optical Papers*, ed. E. Shapiro (Cambridge: Cambridge University Press).

———. 1991. "De motu et sensatione animalium. De vita e morte vegetabili," in M. Mamiani and E. Trucco, "Newton e i fenomeni della vita," *Nuncius*, I, pp. 78–87.

———. 1994. *Trattato sull'Apocalisse*, ed. M. Mamiani (Turin: Bollati Boringhieri).

Rossi, P. 1979. *I segni del tempo: storia della Terra e storia delle nazioni da Hooke a Vico* (Milan: Feltrinelli).

Voltaire. 1962. *Scritti filosofici*, ed. P. Serini, 2 vols (Bari: Laterza).

Westfall, R. S. 1980. *Never at Rest: A Biography of Isaac Newton* (Cambridge: Cambridge University Press).

Whiteside, D. T. 1970. "The Mathematical Principles Underlying Newton's Principia Mathematica," *Journal for the History of Astronomy*, 1, pp. 116–38.

Additional Reading

On the "scientific revolution":
Cohen, I. B. 1985. *Revolution in Science* (Cambridge, MA: Harvard University Press).

Kuhn, T. 1962. *The Structure of Scientific Revolutions* (Chicago: University of Chicago Press).

On the relationship between science and theology
Funkenstein, A. 1986. *Theology and the Scientific Imagination from the Middle Ages to the Seventeenth Century* (Princeton, NJ: Princeton University Press).

Lindeberg, D. C. and Numbers, R. L. (eds). 1986. *God and Nature: Historical Essays on the Encounter between Christianity and Science* (London–Berkeley–Los Angeles: University of California Press).

On the Hermetic tradition and the theme of secrecy
Berselli, L., Poli, C. and Rossi, P. 1983. "Una libera comunità di dilettanti nella Parigi del 600," in P. Rossi et al., *Cultura popolare e cultura dotta nel Seicento* (Milan: Franco Angeli).

Eamon, W. 1994. *Science and the Secrets of Nature: Books of Secrets in Medieval and Early Modern Culture* (Princeton, NJ: Princeton University Press).

Ginzburg, C. 1981. "L'alto e il basso. Il tema della consoscenza proibita nel 500 e 600," *Aut-Aut*, pp. 3–17.

Simmel, G. 1906. "The Sociology of Secrecy and of Secret Societies," *American Journal of Sociology*, XI, pp. 441–98.

Yates, F. A. 1964. *Giordano Bruno and the Hermetic Tradition* (London: Routledge and Kegan Paul).

Zambelli, P. 1994. *L'ambigua natura della magia* (Milan: Il Saggiatore).

On the mechanical arts and technology
Bellone, E. and Rossi, P. (eds). 1982. *Leonardo e l'età della ragione* (Milan: Scientia).

Bloch, M. 1967. *Land and Work in Mediaeval Europe* (New York: Harper & Row).

Borkenau, F. 1934. *Der Übergang vom feudalen zum bürgerlichen Weltbild. Studien zur Geschichte der Philosophie der Manifakturperiode* (Paris).

Cardwell, D. S. L. 1972. *Technology, Science, and History* (London: Heinemann Educational).

Gille, B. 1966. *Engineers in the Renaissance* (Cambridge, MA: MIT Press).

Hall, A. R. 1962. "The Scholar and the Craftsman in the Scientific Revolution," in M. Clagett (ed.), *Critical Problems in the History of Science* (Madison: University of Wisconsin Press), pp. 3–24.

Landes, D. S. 1983. *Revolution in Time* (Cambridge, MA: Belknap Press of Harvard University Press).

Leiss, W. 1972. *The Domination of Nature* (New York: G. Braziller).

Nef, J. U. 1964. *The Conquest of the Material World* (Chicago: University of Chicago Press).

Singer, C., Holmyard, E. J., Hall, A. R., and Williams, T. J. 1954–1978. *A History of Technology*, 7 vols (London: Clarendon Press).

On experimentalism

Shapin, S. and Schaffer, S. 1985. *Leviathan and the Air-Pump* (Princeton, NJ: Princeton University Press).

On illustration, printing, instruments, and the New World

Blunt, W. 1955. *The Art of Botanical Illustration* (London: Collins).

Daumas, M. 1972. *Scientific Instruments of the Seventeenth and Eighteenth Centuries* (New York: Praeger Publishers).

Defossez, L. 1946. *Les Savants du XVIIe siècle et la mesure du temps* (Lausanne).

Dobell, C. 1932. *A. van Leeuwenhoek and His Little Animals* (London).

Eisenstein, E. L. 1979. *The Printing Press as an Agent of Change* (Cambridge: Cambridge University Press).

Gerbi, A. 1973. *The Dispute of the New World* (Pittsburgh, PA: University of Pittsburgh Press).

Gliozzi, G. 1977. *Adamo e il nuovo mondo* (Florence: La Nuova Italia).

Landucci, S. 1972. *I filosofi e i selvaggi* (Bari: Laterza).

Ronchi, V. 1958. *Il cannochiale di Galilei e la scienza del Seicento* (Turin: Einaudi).

On the revolution in astronomy

Burtt, E. A. 1950. *The Metaphysical Foundations of Modern Physical Science* (London: Routledge and Kegan Paul).

Butterfield, H. 1949. *The Origins of Modern Science* (London: Bell and Sons).

Cohen, B. 1974. *The Birth of a New Physics* (New York: Anchor Books).

Crombie, A. C. 1959. *Medieval and Early Modern Science* (New York: Doubleday Anchor Books).

Koestler, A. 1959. *The Sleepwalkers* (London: Hutchinson).

Toulmin, S. and Goodfield, J. 1963. *The Fabric of the Heavens* (London: Penguin).

On Galileo

Banfi, A. 1962. *Galileo Galilei* (Milan: Feltrinelli).

De Santillana, G. 1955. *The Crime of Galileo* (Chicago: University of Chicago Press).

Drake, S. 1978. *Galileo at Work: His Scientific Autobiography* (Chicago: The University of Chicago Press).

Finocchiaro, M. A. 1980. *Galileo and the Art of Reasoning* (Dordrecht: Reidel).

Galluzzi, P. (ed.). 1983. "Novità celesti e crisi del sapere," supplement to the *Annali dell'Istituto e Museo di Storia della Scienza* (Florence).

Geymonat, L. 1965. *Galileo Galilei* (New York: McGraw-Hill).

Redondi, P. 1987. *Galileo Heretic* (Princetion, NJ: Princeton University Press).

Soccorsi, F. 1947. *Il processo di Galilei* (Roma).

Wallace, W. A. 1981. *Prelude to Galileo* (Dordrecht: Reidel).

On Descartes

Grankroger, S. (ed.). 1980. *Descartes: Philosophy, Mathematics, and Physics* (Hassocks: Harvester Press).

Mouy, P. 1934. *Le développement de la physique cartésienne* (Paris).

Nardi, A. 1981. "Descartes e Galilei," *Giornale critico della filosofia italiana*, LX, 1, pp. 129–48.

Pacchi, A. 1973. *Cartesio in Inghilterra* (Rome–Bari: Laterza).

On mechanical philosophy

AA.VV. 1981. *Huygens et la France* (Paris).

Belaval, Y. 1960. *Leibniz critique de Descartes* (Paris: Gallimard).

Casini, P. 1969. *L'universo-macchina* (Bari: Laterza).

Dugas, R. 1950. *Histoire de la mécanique* (Neuchâtel: Griffon).

———. 1954. *La Mécanique au XVIIe siècle* (Paris: Dunod).

Grankroger, S. (ed.). 1980. *Descartes: Philosophy, Mathematics, and Physics* (Hassocks: Harvester Press).

Grant, E. 1981. *Much Ado About Nothing: Theories of Space and Vacuum from the Middle Ages to the Scientific Revolution* (Cambridge: Cambridge University Press).

Heilbron, J. L. 1982. *Elements of Early Modern Physics* (London–Berkeley–Los Angeles: University of California Press).

Kargon, R. II. 1966. *Atomism in England from Hariot to Newton* (Oxford: Clarendon Press).

Lenoble, R. 1934. *Mersenne ou la naissance du mécanisme* (Paris: Vrin).

Mach, E. 1883. *Die Mechanik in ihrer Entwicklung, historisch-kritisch dargestellt* (Leipzig).

Mouy, P. 1934. *Le Développement de la physique cartésienne* (Paris).

Sabra, A. I. 1967. *Theories of Light from Descartes to Newton* (London).

Schofield, R. E. 1970. *Mechanism and Materialism: British Natural Philosophy in an Age of Reason* (Princeton, NJ: Princeton University Press).

On heredity

Jacob, F. 1970. La Logique du vivant: une histoire de l'hérédité (Paris: Gallimard).

On geological time
Gillispie, C. 1959. *Genesis and Geology* (New York: Harper).
Glass, B. (ed.). 1968. *Forerunner of Darwin* (Baltimore, MD: The John Hopkins University Press).
Gould, S. J. 1987. *Time's Arrow, Time's Cycle: Myth and Metaphor in the Discovery of Geological Time* (Cambridge, MA: Harvard University Press).
Greene, J. C. 1959. *The Death of Adam* (Arnes: Iowa State University Press).
Morello, N. 1979. *La nascità della paleontologia nel Seicento: Colonna, Stenone e Scilla* (Milan: Angeli).
Rossi, P. 1969. *Le sterminate antichità: studi vichiani* (Pisa: Nistri-Lischi).
Toulmin, S. and Goodfield, S. 1965. *The Discovery of Time* (London: Hutchinson).

On classification
Barsanti, G. 1992. *La scala, la mappa, l'albero. Immagini e classificazioni della natura fra Sei e Ottocento* (Florence: Sansoni).
Slaughter, M. 1982. *Universal Languages and Scientific Taxonomy in the Seventeenth Century* (Cambridge: Cambridge University Press).

On the calculus
Bottazzini, U. 1988. "Curve e equazioni, i logaritmi, il calcolo," in P. Rossi (ed.), *Storia della scienza moderna e contemporanea*, vol. I (Turin: Utet), pp. 261–320, 449–84.
Boyer, C. B. 1949. *The History of the Calculus and Its Conceptual Development* (New York: Dover).
Fraiese, A. 1964. *Galileo matematico* (Rome: Editrice Studium).
Geymonat, L. 1947. *Storia e filosofia dell'analisi infinitesimale* (Turin: Levrotto and Bella).
Kline, M. 1972. *Mathematical Thought from Ancient to Modern Times* (New York: Oxford University Press).
Petitot, J. 1979. "Infinitesimale," *Enciclopedia Einaudi*, vol. VII (Turin: Einaudi), pp. 443–57.
Whiteside, S. 1961. "Patterns of Mathematical Thought in the Later Seventeenth Century," *Archive for the History of the Exact Sciences*, I, pp. 179–388.
Zeldovich, Y. B. 1973. *Higher Mathematics for Beginners* (Moscow: Mir).

On Newton
Casini, P. 1969. *L'universo-macchina* (Bari: Laterza).
Cohen, I. B. 1956. *Franklin and Newton* (Philadelphia: American Philosophical Society).
Schofield, R. E. 1970. *Mechanism and Materialism: British Natural Philosophy in an Age of Reason* (Princeton, NJ: Princeton University Press).
Vavilov, S. I. 1943. *I. Newton* (Moscow: Akademija Nauk).
Westfall, R. S. 1958. *Science and Religion in Seventeenth-Century England* (New Haven, CT: Yale University Press).
——. 1971. *Force in Newton's Physics* (London: Macdonald).

Index

Abbri, Ferdinando, 143, 147
About the Excellency and Grounds of the Mechanical Hypothesis (Boyle), 133
Academia Secretorum Naturae, 195
Académie Royale des Sciences, 198–9
academics, 1, 23–4, 194–202
Academy of Montmor, 197
Accademia degli Inquieti, 202
Accademia dei Lincei, 24, 51, 77, 195–6
Accademia del Cimento, 196–7
Accademia Naturae Curiosorum, 200
acceleration, 10, 89, 91, 95, 96–8
Acosta, José, 53, 54, 55
d'Acquapendente, Girolamo Fabrici, 131, 157
Acta Eruditorum, 171, 202, 214
activity, and substance, 137
Ad Vitelionem paralipomena (Kepler), 68
Adam, 22
Adelmann, H. D., 160
aemulatio, 42, 43
Agricola, Georg, 16, 23, 31, 36, 38
Agrippa, Cornelius, 21, 22
air
  Aristotelian philosophy, 12
  Boyle, 145
  Descartes, 104, 184

  Helmont, 143
  Paracelsus, 141
  Peripatetics, 184
  Roberval, 184
air pressure, 184–5
Alberti, Leon Battista, 31, 32
alchemy, 20–1, 139, 142
  Agricola, 36
  Aristotle, 19
  Bacon's doctrine of forms, 26
  Biringuccio, 23, 35–6
  Mersenne, 24
  Newton, 7, 26, 215, 224–5
  Paracelsus, 140, 141
Aldrovandi, Ulysses, 47, 181
D'Alembert, Jean, 39
Alexander the Great, 19
Alexandria, 13
algebra, 102, 188
alienation, 40
allusion, 21
Almagest (Ptolemy), 13, 42, 56, 59, 74
Altieri Biagi, M. L., 196, 197
Ammannati, Giulia, 73
Ammonius, 22
analogy, 21, 123–4, 125–6
Anatomia (de Luzzi), 45
anatomical texts, 44–6
anatomy, 45–6, 157–63
Anatomy of the World (Donne), 63

Anaximander, 223
d'Andrea, Francesco, 197
De anima (Aristotle), 192
animal classification, 175–6, 181, 182
animalculism, 160–3
animals
    changes over time, 166–7, 173
    extinction, 166–7, 168–9
    mechanical philosophy, 129–32
    New World, 53–5
animism, 20, 62
Antal, F., 32
anthropocentrism, 114, 120–1
anthropomorphism, 20, 126
antipathy, 27, 28, 148, 149
De antiquissima (Vico), 129
antiquity, 42–3, 53
Apocalypse, 228
Apollo, 168, 223
Apollonius, 21, 42
Apologia pro Galilaeo (Campanella),
        109–10, 116
Archidoxis (Paracelsus), 141
Archimedes, 31, 33, 42, 139
    and Galileo, 73, 74, 85–6
    as a mechanic, 16
    and Stevinus, 37
architecture, 31, 32
Areopagitica (Milton), 94
Argumenta contra Galilaeum, 109
Ariosto, 115
Aristarchus, 223
Aristotle
    De anima, 192
    De generatione et corruptione, 192
    Parva naturalia, 192
Aristotle/Aristotelian philosophy, 51
    bodies resistant to penetration, 137
    color, 211
    comets, 64
    cosmology, 12–13, 87–9, 110, 151
    Earth, 13, 151, 166
    elements, 12
    floghistòs, 147
    force, 10
    fossils, 166
    and Galileo, 73–4, 85–6, 87–9, 110

and Gassendi, 127
and Gilbert, 151
infinity, 186
and Leibniz, 137
mechanics, 15
Meteorologica, 166
motion, 12–13, 73–4, 89, 90, 184
and Petrus Ramus, 44
Physics, 184, 192
planetary motion, 13, 159
Posterior Analytics, 192
Secreta secretorum, 18–19
soul, 151
terrestrial world, 12
theory of types, 128
velocity, 10
voids, 73–4, 104, 137, 184
arithmetic, 15, 102, 129, 188
Arius, 226
Arnaldi, G., 194
Arnoldo, Brother, 23
art, 127–8
artists, 32–3
Ash Wednesday Supper (Bruno), 108
The Assayer (Galileo), 83–5, 93
astral body, 26
astrobiology, 109
astrology, 14, 20, 139
    Aristotle, 19
    Brahe, 26
    illnesses, 28
    Paracelsus, 140, 141
Astronomia nova seu Physica coelestis
        (Kepler), 3, 68, 69–70, 71, 87
Astronomiae instaurate mechanica
        (Brahe), 64
astronomy, 43–4, 56–72, 139
    Galileo, 48–50, 74–83, 87–9
    Hermetic-Paracelsian tradition, 140
    as liberal art, 15
    see also Brahe, Tycho; Copernicus,
        Nicolaus; Kepler, Johannes; moon;
        motion, planetary; planets;
        Ptolemy; stars; sun
Athanasius, 226
atheism, 132, 173
atmosphere, 145

atmospheric pressure, 184–5
atomism, 84, 104, 117, 132, 133
attraction
    Fracastoro, 27
    magnetism, 148, 149, 150–1, 154
    Newtonian physics, 216–17
De augmentis (Bacon), 128
Averlino, Francesco, 31
Averroism, 19, 117
Avicenna, 140, 192

Bach, Johann Sebastian, 2
Bachelard, Gaston, 9, 10
Bacon, Francis, 1, 3, 27, 38, 99
    ancient civilizations, 221
    ancient legends, 222
    antiquity, 43
    Aristotelian theory of types, 128
    De augmentis, 128
    Copernicanism, 62
    and Descartes, 101
    doctrine of forms, 26
    empirical knowledge, 35
    Galileo's discoveries, 24, 62
    industry of mechanics, 49
    inventions, 24, 41
    and Kepler, 72
    man as master of nature, 35, 40
    manual arts, 16
    memory, 180
    mistrust of hypotheses, 203
    nationalism, 201
    nature, 35, 40, 166, 183–4
    New Atlantis, 51
    and Newton, 107
    Novum Organum, 3, 44, 227
    objective/subjective qualities of
        bodies, 125
    ritual magic, 24
    scientific instruments, 183–4
    scientific method, 24
    technical knowledge, 40
    universal language, 177
    universities, 193
Bacon, Roger, 19
bacteria, 53
Bahuin, Gaspar, 180

Baillet, Adrien, 3
banausia, 15
Barbaro, Daniele, 31, 36
Barberini, Maffeo *see* Urban VIII, Pope
barometer, 184–5, 186
Barrow, Isaac, 210, 213
Basilica chymica (Croll), 142
Battle of Anghiari (Leonardo da Vinci),
    34, 44
Bayle, François, 154
Bayle, Pierre, 107
Becher, Joachim, 146–7
Beeckman, Isaac, 100, 193–4
bees, 51
Beghinselen der Weeghconst (Stevinus),
    37
Beguin, Jean, 143
Being, 50
Bellarmine, Cardinal Robert, 79, 81,
    82, 83, 93
Belon, Pierre, 47
Ben David, J., 195
Benedetti, Giovanni Battista, 61
Bentley, Richard, 173
Berengario de Carpi, Jacopo, 45
Bergerac, Cyrano de, 117, 120
Berkeley, George, 190–1
Berlin, 200–1
Bernardi, W., 160, 163
Bertrand, J., 198
Besson, Jacques, 31
Bevilacqua, F., 211
Bianchi, L., 5, 6, 219
Bible, 54, 60
    age of the Earth, 167, 168, 169–70
    Castelli, 81
    Galileo, 78–80
    Newton, 220–1, 225, 226–8
Bibliotheca universalis (Gesner), 47
La bilancetta (Galileo), 73, 74
binomial rule, 213
Biringuccio, Vanoccio, 23, 31, 35–6, 38
The Black Cloud (Hoyle), 121
blood cells, 53
blood circulation, 45–6, 157–60
Blundeville, Thomas, 61
Bologna, 201–2

Bono of Ferrara, 21
*Book of Diverse Speculation on
   Mathematics and Physics*
   (Benedetti), 61
books, 25, 41–3
   illustrations, 44–8, 52
Borel, Pierre, 43–4, 51, 117, 120
Borelli, Giovanni Alfonso, 72, 130–1,
   160, 196
Borgia, Cesare, 33
Borromini, Francesco, 2
Borselli, L., 198
botanical classification, 175–82
botanical gardens, 47
botanical texts, 44, 47
Boyle Lectures, 173
Boyle, Robert
   *About the Excellency and Grounds
      of the Mechanical Hypothesis*, 133
   chemical elements theory, 143, 144–5
   *Considerations Touching the
      Usefulness of Experimental
      Natural Philosophy*, 38–9
   mechanical philosophy, 126, 133,
      134, 144–5
   micro-entities, 122
   nature, 40, 133, 134
   New Experiments, 210
   and Newton, 107, 213
   Royal Society, 199, 200
   The Sceptical Chymist, 144
   voids/atmospheric pressure, 185
Boyle's Law, 145
Bradwardine, Thomas, 5
Brahe, Tycho, 1, 61, 88
   astrology, 26
   Astronomiae instaurate mechanica,
      64
   comets, 64, 83
   cosmology, 12, 63–5
   Earth, 65, 75
   and Galileo, 83, 86
   infinity, 110, 115
   and Kepler, 64–5, 67, 68, 69
   multiplicity of worlds, 110
   *On the Most Recent Phenomena of
      the Aetherial World*, 65

On the New Star, 64
   orbit of Mars, 68
   stars, 64, 65, 75
   Uraniborg observatory, 64, 194
brain, 157
"A Brief Demonstration of a
   Memorable Error by Descartes"
   (Leibniz), 136
*A Brief instruction in military
   architecture* (Galileo), 74
*A Briefe and Troue Report of the New
   Found Land* (Hariot), 54
Brizio, A. M., 34
Brunelleschi, Filippo, 31
Brunfels, Otto, 47, 180
Bruni, Leonardo, 42
Bruno, Giordano, 27, 69
   Ash Wednesday Supper, 108
   and Borel, 117
   Copernicanism, 61, 62, 108
   death of, 3, 80, 108
   Earth, 62, 108
   and Galileo, 81, 113
   infinity, 108–9, 110, 111, 112, 115,
      120
   De l'infito, universo e mondi, 62,
      108–9
   and Kepler, 111
   life on other planets, 118
   and Lucretius, 110, 184
   magic, 21
   mechanics, 16
   multiple worlds, 110
   New World, 54
   stars, 109, 119
   works banned, 80
Buffon, Georges Louis Leclerc, Comte
   de, 55, 164, 171
Bureau d'Adresse, 197–8
Buridan, Jean, 5
Burnet, Thomas, 120, 169–70, 171,
   173

Cabal, 26, 114, 117
Cabeo, Nicholas, 148, 151
Cabinet des frères Dupuy, 197
Caccini, Tommaso, 60, 80–1, 82

Calcar, Jan Stephan van, 45
calcination, 145, 147
calculation, 183
calculus, 97, 163
   Leibniz, 188, 189–91, 204, 214
   Newton, 7, 188–91, 203, 204, 213
Callicles, 15, 16
Calvin, John, 60
Camerarius, Joachim, 43
Campanella, Tommaso, 27, 99
   Apologia pro Galilaeo, 109–10, 116
   and Bergerac, 117
   and Borel, 117
   The City of the Sun, 41
   freed by Urban VIII, 86
   imprisonment, 80
   magic, 21
   mechanical inventions, 41
   multiplicity of worlds, 109–10
   universal theocracy, 40
   works banned, 80
Camporeale, S., 60
capacity, 149
capillarity, 122, 124
Caravaggio, 2
Cardano, Gerolamo, 27, 54, 188
Casini, P., 226
Caspar, M., 2
Cassini, Gian Domenico, 194, 198
Cassiopeia, 64
Cassirer, Ernst, 20, 102–3
Castagno, Andrea del, 32
Castelli, Benedetto, 77, 81
Castelnuovo, G., 189
The Castle of Knowledge (Recorde), 61
catastrophism, 165
causes, analogy of, 123–4
De causis corruptarum artium (Vives),
   30
De causis criticorum diebus
   (Fracastoro), 28
Cavalieri, Bonaventura, 188, 189, 190,
   202
Cavendish, Lord Henry, 149
Cavendish, Margaret, 116
celestial world *see* heavenly bodies
cells, 52

Centuria observationum
   microscopicarum (Borel), 51
Cesalpino, Andrea, 54, 181
Cesi, Prince Federico, 52, 54, 75, 78
   Accademia dei Lincei, 51, 195, 196
Chambers's encyclopedia, 39
charge, 149
charlatans, 28
Charles I, King of England, 158
Charles II, King of England, 199
Charles V, Emperor, 32, 36, 45
chemical change, 122
chemistry, 130, 139–47
Christianity, 132–5, 225–6
*Chronology of Ancient Kingdoms
   Amended* (Newton), 220
Church Fathers, 14, 53, 79, 226
Cicero, 184, 226
circulatory systems, 45–6, 157–60
The City of the Sun (Campanella), 41
civil philosophy, 128
Clagett, Marshall, 5
Clarke, Dr. Samuel, 137, 218
classicism, 42–3, 53
classification, 175–82
Clavelin, M., 92
Clavis totius philosophiae chymicae
   (Dorn), 142
Clavius, Christoph, 101
Clement VII, Pope, 81
Clement VIII, Pope, 61, 80
coherence, 84
Colbert, Jean-Baptiste, 198
*A Collection of English Words not
   Generally Used* (Ray), 178
collision theory, 106
Colonna, Fabio, 168
color, 211–12
Columbus, Christopher, 49, 53, 116
Columbus, Realdo, 157
combustion, 10, 124, 145, 147
Comenius, Amos, 25, 40, 199
comets, 64, 83, 105, 207, 220
Commandino, Federico, 42
commercialization, 40
common sense, 9, 10
compass, 41, 74, 150

Compendium (Gohory), 142
Compendium musicae (Descartes), 100
conatus, 136
condensation, 124
conductors, 148
Confucius, 226
Conics (Apollonius), 42
*Considerations Touching the
    Usefulness of Experimental
    Natural Philosophy* (Boyle), 38–9
De contagionibus et contagiosis morbis
    (Fracastoro), 27
contagions, 27
continuity, 186–7, 190
coordinates, 102
Copernicus, Nicolaus, 1, 12, 44, 45,
    56–63, 65, 203, 224
  and Bergerac, 117
  and Borel, 117
  and Bruno, 61, 62, 108
  celestial orbits, 67
  condemnation of, 82–3
  Earth, 56, 57, 58, 59, 60, 74–5, 82
  and Galileo, 61, 63, 74, 75, 79–83,
    87, 93
  heliocentricity, 26
  *De hypothesibus motuum coelestium
    commentariolus*, 57
  infinity, 108, 109, 110
  and Kepler, 56, 61, 63, 66–7, 69, 81
  multiplicity of worlds, 108, 110
  planets, 56, 57–60
  positive portrayal of, 216
  *De revolutionibus orbium
    coelestium*, 2, 56, 57–9, 60–1, 81,
    108
  sun, 57, 58–9, 82
Corneille, Pierre, 2
Cornelio, Tommaso, 99, 197
De corpore (Hobbes), 24
Corpus Hermeticum, 19, 26
Cosimo II, 49
Cosimo III, 153
cosmographia, 193
The Cosmographical Mystery (Kepler),
    61, 66, 67–8, 70, 71
cosmology, 12–14, 56–72, 164–5

*Cosmotheoros sive de terris coelestibus
    earumque ornatu conjecturae*
    (Huygens), 118, 121
Coulomb, Charles, 149
Cours de chimie (Leméry), 146
Craterius, 23
creation, 134–5, 142
Cremonini, Cesare, 50, 76, 77, 80
Croll, Oswald, 142
Cromwell, Oliver, 193
crustaceans, 52
cultural reform, 21

Daedalus, 40
Dalgarno, George, 177
Dallari, U., 193
Daniel, 227
Daniel, Gabriel, 130
Dante Alighieri, 14
Darwin, Charles, 59, 216
De Libera, A., 4
De Martino, Ernesto, 155–6
Debus, A. G., 143, 144
decimal fractions, 37
Dee, John, 62
*Defense against the deception and fraud
    of Baldessar Capra* (Galileo), 74
Defoe, Daniel, 3
deism, 226
*Dell'anatomia et delle infermitadi del
    cavallo* (Ruini), 47
Delle fortificationi (Lorini), 36–7
Democritus, 27, 109–10, 117, 132, 137
derivation, 189
Descartes, René, 5, 25, 27, 38, 71, 84,
    99–107, 165
  air, 104, 184
  antiquity, 43
  art and nature, 128
  bodies resistant to penetration, 137
  color, 211
  Compendium musicae, 100
  cosmology, 102–6
  creation, 134, 170
  Dioptrique, 101, 123, 210
  Discourse on Method, 24, 101, 129
  Earth, 105, 154, 168, 169

Epistola ad Gilbertum Voetium, 101
and Gassendi, 127, 134
Géométrie, 101, 102, 189, 213
geometry, 102, 104, 106–7
God, 103, 104, 114, 133, 135, 138
Harvey's blood circulation theory,
  158
Hermetic tradition, 26
inertia, 91, 103
infinity, 113–14
and Kepler, 72, 101, 106
and Kircher, 155
language, 178
and Leibniz, 99, 135, 136
magnetism, 153–4
manual arts, 16
mathematics, 102–3, 107, 188
matter, 103–5, 133, 134, 135, 184
mechanical philosophy, 123, 126,
  129–30, 137, 209
Meditationes de prima philosophia,
  101, 127
and Mersenne, 72, 94, 106
Météores, 101, 104–5
micro-entities, 122
motion, 11–12, 91, 103, 104–6, 133,
  136
nature, 40, 103, 106–7, 123, 134,
  135
Neapolitan Accademia degli
  Investiganti, 197
and Newton, 106, 107, 134, 189,
  203, 206, 213
objective/subjective qualities of
  bodies, 125
physics, 11–12, 102–7, 130, 137,
  203, 204, 209
physiology, 129–30, 160
positive portrayal of, 216
Principles, 11–12, 101, 106, 107,
  113, 134, 135, 153
religious beliefs, 225
reproduction, 162
Rules for the Direction of the Mind,
  100, 223
and Scilla, 168
theory of vortices, 105–6, 206

Traité des passions de l'âme, 101
Treatise on Man, 126, 128
universities, 193
voids, 104, 105, 137, 184
The World, 101, 103, 134
"world-maker," 173
Description of a New World
  (Cavendish), 116
diabolicentric theory, 116
*Dialogue on the Two Chief World
  Systems* (Galilco), 82, 85, 86–93,
  113, 202, 213
Dictionariolum trilingue (Ray), 178
Dictionnaire français (Richelet), 16
Diderot, Denis, 39–40, 132
differentials, 190
Digby, Kenelm, 117
digestion, 143
Digges, Thomas, 61, 62, 115
Dijksterhuis, E. J., 124–5, 150, 185–6
Diogenes, 184
Dioptrice (Kepler), 70, 210
Dioptrique (Descartes), 101, 123, 210
Dioscorides, 179
Discours admirables (Palissy), 29
*Discours nouveau prouvant que les
  astres sont des terres habitées*
  (Borel), 117
Discours sur la cause de la pesanteur
  (Huygens), 107
*Discours touchant la methode de la
  certitude et l'art d'inventer*
  (Leibniz), 39
discourse, and practice, 29–30, 36
Discourse on comets (Guiducci), 83
Discourse on Method (Descartes), 24,
  101, 129
Discourse on Science and Art
  (Rousseau), 4
Discourse on the tides (Galileo), 82
Discourses on Earthquakes (Hooke),
  164, 166
Discourses on Two New Sciences
  (Galileo), 3, 37, 44, 84, 95–8, 131,
  187
Discovery of a New World (Wilkins),
  116

dissection, 44
*De dissectione partium corporis
    humani* (Estienne), 45
Dissertatio cum Nuncio Sidereo
    (Kepler), 70, 111
distillation, 144
Diverse et artificiose machine (Ramelli),
    31, 37–8
Divine Comedy (Dante), 14
Dobell, C., 162
Donne, John, 63
Dorn, Gérard, 142
The Dream (Kepler), 115–16, 121
druggists, 139
Dryden, John, 44
Dubois, Jacques, 46
Duchesne, Joseph, 142
Duhem, Pierre, 5
Dürer, Albrecht, 44, 48, 180
    engraving, 45
    rhinoceros head, 47, 181
    treatise on descriptive geometry, 31
    treatise on fortifications, 31
    *Treatise on the proportions of the
        Human Body*, 38
Dury, John, 193
dynamics Leibniz, 136, 137
    mechanical philosophy, 122, 124
    see also force; motion

Eamon, W., 19, 23
Earth
    Aristotelian philosophy, 13, 151, 166
    Brahe, 65, 75
    Bruno, 62, 108
    Burnet, 169–70
    Cabeo, 151
    centrality and immobility of, 13, 14,
        57, 65
    Copernicus, 56, 57, 58, 59, 60, 74–5,
        82
    Descartes, 105, 154, 168, 169
    Galileo, 88–90, 92
    Gilbert, 150, 151, 152
    Guericke, 154–5
    Hooke, 164, 166–7
    Kepler, 67, 68, 111, 112

Kircher, 168
Leibniz, 171–3
Newton, 207, 219
Patrizi, 61–2
Scilla, 167–8
Temple, 170
Whiston, 170–1
Wilkins, 116
Woodward, 170
    see also geology
earth (element)
    Aristotelian philosophy, 12
    Descartes, 104
    Paracelsus, 141
earthquakes, 166
Edison, Thomas, 9
effects, analogy of, 123–4
Egyptians, ancient, 221, 223
elasticity, 124, 145
electric field, 149
electricity, 148–9, 151, 155
electrology, 149
electroscope, 151
elements
    Aristotelian philosophy, 12
    Descartes, 104–5
    Helmont, 143
    Paracelsus, 141
*Eléments de botanique ou méthode
    pour reconnaître les plantes*
    (Tournefort), 179
Elements (Euclid), 74, 203
Elements of Physics (Melanchthon), 60
Elias, Brother, 23
Elizabeth, Queen of Sweden, 101
Emery, C., 178
encyclopedia, 177–8
Encyclopédie, 4, 132
*Encyclopédie ou dictionnaire raisonné
    des sciences, des arts et des
    mestiers*, 39
engineers, 29–40
England, universities, 192, 193
engravings, 44, 45, 52
Entretiens sur la pluralité des mondes
    (Fontenelle), 117
Epicureanism, 15, 133

Epicurus, 27, 84, 109–10, 117
epistemological obstacles, 9
Epistola ad Gilbertum Voetium
    (Descartes), 101
Epitome of Astronomy (Maestlin), 61
Epitome of Copernican Astronomy
    (Kepler), 70
Epitome (Vesalius), 45
Erasmus of Rotterdam, 30, 36, 42
*Esperienze intorno alla generazione
    degli insetti* (Redi), 161
Essay de dynamique (Leibniz), 137
*Essay Towards a Natural History of
    the Earth* (Woodward), 170
*An Essay Towards a Real Character
    and a Philosophical Language*
    (Wilkins), 179
Essays (Montaigne), 55
Estienne, Charles, 45
etchings, 45
ethics, 140
Eucharist, 84
Euclid, 31, 33, 42, 63, 139
    in *Discourses on Two New Sciences*,
        95
    Elements, 74, 203
    and Newton, 203, 204
    polyhedrons, 119
    theory of proportions, 95
Eudoxus of Cnidus, 13
Eustachio, Bartolomeo, 157
Eve, 163
evil, 40
*Examination of Dr. Burnet's Theory of
    the Earth* (Keill), 173
experience, 5, 122, 125
Experiences nouvelles touchant le vide
    (Pascal), 185
Experimenta nova (Guericke), 154
Experimental Philosophy (Power),
    50–1
experimentation, 5, 183
De expetendis et fugiendis rebus, 115

Faber, Johannes, 49, 51
Fabri, Honoré, 137
De Fabrica (Vesalius), 30, 45–6

Fabricius of Acquapendente, 131, 157
Falloppius, Gabriel, 157
Fardella, Michelangelo, 99
Farrar, W. V., 201
fate, 134
Febvre, L., 42
Ferdinand I, Grand Duke, 73
Ferdinand II, Grand Duke, 196
Ferdinand, King of Austria, 36
Fermat, Pierre, 188
Fernel, Jean, 159–60
Feynman, R. T., 190, 212
Ficino, Marsilio, 19, 20, 26, 62, 153,
    159
Filopono, Giovanni, 81
fire
    Aristotelian philosophy, 12
    Boyle, 145
    Galileo, 84
    Glauber, 144
    Helmont, 143
    Paracelsus, 141
*Five Books of the Harmonies of the
    World* (Kepler), 2, 70
flies, 52
floghistòs, 147
Fludd, Robert, 72, 142
fluents, 189
fluid dynamics, 206
fluxions, 188, 189, 213, 214
Fontana, Felice, 51
Fontana, Niccolò, 31
Fontenelle, Bernard le Bovier de, 116,
    117, 118, 120, 170, 180
force
    ancient physics, 11
    Aristotle, 10
    Descartes, 11–12
    Galileo, 89
    Kepler, 67
    Newton, 10, 204, 205
    see also gravity
forms, Bacon's doctrine, 26
Foscarini, Paolo Antonio, 81, 82
fossils, 165–9, 170, 172–3
Fracastoro, Girolamo, 27–8
France, universities, 192, 193

Francis I of France, 34
Frederick I of Prussia, 201
Frederick II of Prussia, 201
Freud, Sigmund, 20
friction, 151
Fuchs, Leonhart, 43, 47
De fundamentis astrologiae certioribus
	(Kepler), 68
Furni novi philosophici oder
	Beschreibung einer neue erfunden
	Distillirkunst (Glauber), 144

Galen, Claudios, 43, 192
	botany, 179
	circulation system, 45–6, 157–8, 159,
		160
	On the Use of Parts, 44
	and Paracelsus, 140
Galilei, Vincenzio, 73
Galileo, 1, 5–6, 35, 39, 53, 71, 73–98,
		99, 203
	Accademia dei Lincei, 195
	Accademia del Cimento, 196, 197
	and Archimedes, 73, 74, 85–6
	and Aristotle, 73–4, 85–6, 87–9, 110
	The Assayer, 83–5, 93
	astronomy, 48–50, 74–83, 87–9
	Bacon's praise of, 24, 62
	and Bergerac, 117
	Bible, 78–80
	La bilancetta, 73, 74
	Book of Scripture/Book of Nature,
		228
	and Brahe, 83, 86
	A Brief instruction in military
		architecture, 74
	and Bruno, 81, 113
	condemnation by the Inquisition, 3,
		93–4, 101
	Copernicanism, 61, 63, 74, 75,
		79–83, 87, 93
	Defense against the deception and
		fraud of Baldessar Capra, 74
	Dialogue on the Two Chief World
		Systems, 82, 85, 86–93, 113, 202,
		213
	Discourse on the tides, 82

Discourses on Two New Sciences, 3,
		37, 44, 84, 95–8, 131, 187
	Earth, 88–90, 92
	Geometrical and Military Compass,
		74
	and Hariot, 54
	infinity, 113, 115, 187–8
	and Kepler, 50, 67–8, 72, 74, 92–3,
		111–12, 113
	Letter on Sunspots, 78–9
	Letter to Castelli, 78, 79, 81
	Letter to Liceti, 113
	Letter to Madame Christina of
		Lorraine, 82, 202
	magnetism, 149
	mathematics, 12, 85, 102
	Mechaniche, 95
	mechanics, 124
	micro-entities, 122
	moon, 48–9, 92–3
	motion, 73–4, 83–4, 88–93, 95,
		96–8, 103, 106
	De Motu, 73, 95, 96
	multiplicity of worlds, 109–10
	nature, 40, 78–9, 84–5, 106
	Neapolitan Accademia degli
		Investiganti, 197
	and Newton, 213
	objective/subjective qualities of
		bodies, 125
	On Fortifications, 74
	On Mechanics, 74
	Opere, 202
	physics, 10, 12, 95–8, 102
	planets, 49, 75–6
	Platonism, 85–6
	positive portrayal of, 216
	relativity, 88–90
	and Scilla, 168
	Sidereus nuncius, 48, 49, 74, 111
	stars, 48, 49, 113
	sun, 79–80, 87
	telescope, 17, 48–50
	Treatise on the Sphere, or
		Cosmography, 74
	velocity, 73, 96–8
Gallino, L., 183

Galluzzi, P., 196, 197
Gamba, Marina, 76
Garin, E., 60
gases, 143
Gassendi, Pierre, 99
  Aristotelianism, 127
  and Bergerac, 117
  bodies resistant to penetration, 137
  and Descartes, 127, 134
  and Fludd, 142
  God, 127, 132
  imagination, 123
  micro-entities, 122
  natural/artificial objects, 127
  objective/subjective qualities of
    bodies, 125
  Syntagma philosophicum, 127
  voids, 137
Geber, 23
gems, 149
"General Scholium" (Newton), 134,
  204–5, 208–9
generation, 160–3
De generatione animalium (Harvey),
  161
De generatione et corruptione
  (Aristotle), 192
Gentile theology, 222
Geography (Ptolemy), 42
geology, 36, 139, 164–74
*Geology or a Discourse Concerning the
  Earth Before the Deluge* (Temple),
  170
Geometrical and Military Compass
  (Galileo), 74
Géométrie (Descartes), 101, 102, 189,
  213
geometry
  comparison of infinites, 188
  Descartes, 102, 104, 106–7
  exhaustion method, 187
  Hobbes, 128
  as liberal art, 15
  Newton, 188–9, 203
  as solution to arithmetic/algebraic
    problems, 188
  Vico, 129

Gesner, Konrad, 47, 181
Ghiberti, Lorenzo, 31, 33
Gilbert, William, 24, 27, 62, 63
  infinity, 115
  and Kepler, 69
  De magnete, 115, 149–50
  magnetism, 26, 29, 149–51, 152
  *De mundo nostro sublunari
    philosophia nova*, 44
Giorello, G., 189
Giusti, E., 188, 191
Glanvill, Joseph, 213
Glauber, Rudolph, 144
Gmelin, Johan Friedrich, 180
Gnosticism, 19
God, 6, 20, 22
  Burnet's sacred theory of the Earth,
    169–70
  confusion of languages as
    punishment, 177
  creation of mountains, 168
  creation of reason, 129
  Descartes, 103, 104, 114, 133, 135,
    138
  Gassendi, 127, 132
  knowledge of final causes and
    essences, 127, 128
  Leibniz, 135–6, 138, 171, 218
  mechanical philosophy, 132–5
  and Moses, 221
  Newton, 208, 217, 218, 225–7,
    228
  Paracelsus, 141
  Spinoza, 132
Godwin, Francis, 116
Gohory, Jacques, 142
Gombrich, E. H., 47, 48
Gorgias (Plato), 15
Gould, S. J., 176
Graaf, Reinier de, 162
grammar, 15
Grant, E., 184
Grassi, Orazio, 83, 84
gravity
  Copernicus, 57
  Newton, 10, 204, 207–9, 213, 217,
    223

The Great Chain of Being (Lovejoy),
    110
Greenwich Royal Observatory, 200
Grew, Nehemia, 51
Grimaldi, Francesco Maria, 210
Gualdo, Paolo, 77
Guericke, Otto van, 154–5, 185
Guiducci, Mario, 83
gunpowder, 41
Gutenberg, Johannes, 41

Habert of Montmor, Henri Louis,
    197
Hacking, I., 183
Hackmann, W. D., 194
Hahn, Roger, 199
Hall, A. R., 52, 201
    Accademia del Cimento, 196
    Leibniz and Newton quarrel, 188,
        214
    Royal Society, 199, 200
Hall, John, 23, 193
Halley, Edmund, 112, 171
Hammerstein, N., 201
Hariot, Thomas, 31, 49, 54
Hartlib, Samuel, 25, 199
Hartsoeker, Nicolaus, 146, 163
Harvey, Gabriel, 101, 131–2
Harvey, William, 1, 26, 39, 158–60,
    161, 162
De Havenvinding (Stevinus), 37
hearing, 123
heart, 26, 157–60
heat, 83–4, 122
heavenly bodies
    distinction from terrestrial world,
        12–14
    see also astronomy; moon; planets;
        stars; sun
Hegel, Georg Wilhelm Friedrich, 55
Heilbron, J. L., 149, 154, 155
Helmont, Jean-Baptiste van, 142–3,
    145
herbals, 47
Herbarum vivae icones (Brunfels), 47,
    180
Hermes Trismegistus, 19, 21, 23, 26, 62

Hermetic Platonism, 153
Hermeticism, 6, 18–22, 26–8, 37
    ancient kingdoms, 220, 221, 223
    chemical philosophy, 140
    De Martino, 156
    infinity, 115
    Kepler, 26, 71
    mechanical philosophy, 132, 133
Hernández, Francisco, 54
Hero, 31
Herwart of Hohenburg, 70
Hippocrates, 42–3, 192
Histoire comique des états et empires de
    la Lune (Bergerac), 117
L'histoire de la nature des oyseaux
    (Belon), 47
Histoire naturelle (Buffon), 171
histology, 51
Historia animalium (Gesner), 47, 181
Historia general natural de las Indias
    (Oviedo), 53–4
Historia natural y moral de las Indias
    (Acosta), 54
Historia Plantarum (Ray), 178
De historia stirpium (Fuchs), 47
historical continuity, 9
history, 129, 164–5
History of animals (Gesner), 47, 181
History and Demonstration Concerning
    Sunspots and their Phenomena, 75,
    78
Hobbes, Thomas
    all is possible, 135
    all psychic life as mechanical
        philosophy, 132
    De corpore, 24
    and Descartes, 99
    Harvey's blood circulation theory,
        158
    Leviathan, 125, 193
    mathematical digressions, 27
    and Newton, 213
    objective/subjective qualities of
        bodies, 125, 126
    politics as mechanical philosophy,
        127
    truths in physics, 128

universities, 193
d'Holbach, Baron, 120
Hooke, Robert, 165
    analogies, 123–4
    color, 211
    Earth, 164, 166–7
    fossils, 166
    light, 210
    micro-entities, 122
    Micrographia, 123, 166, 210
    microscope, 51–3, 123, 160
    Microscopia, 52
    natural history, 166–7
    and Newton, 107, 213, 214
    Royal Society, 200
Horky, Martin, 50
Horsley, Samuel, 215
Hoyle, Fred, 121
Hubble telescope, 183
Hues, Robert, 31
De humani corporis fabrica (Vesalius),
    30, 45, 46
humanism, 42
Huygens, Christiaan, 1, 117, 118–20,
    155
    Académie Royale des Sciences, 198,
        199
    centrifugal force, 204
    *Cosmotheoros sive de terris
        coelestibus earumque ornatu
        conjecturae*, 118, 121
    and Descartes, 106, 107, 201
    Discours sur la cause de la pesanteur,
        107
    and Kepler, 118–19
    light, 210
    mathematics, 12, 102
    micro-entities, 122
    De motu corporum ex percussione,
        106
    and Newton, 214
    physics, 12, 137
    Traité de la lumière, 107
    university study, 193
hydrochloric acid, 144
hydrostatic balance, 73
Hypomnemata mathematica

(Stevinus), 37
*De hypothesibus motuum coelestium
    commentariolus* (Copernicus), 57

Ianniello, M. G., 211
iatrochemistry, 139, 140, 143–4, 146
iatromechanics, 130
Idea medicinae philosophicae
    (Severinus), 142
Ideas on the uncertainty of medicine
    (Leonardo di Capua), 99
Iliastrum, 141
illnesses, 28
illumination, 9–10
illustrations, 44–8, 52
imagination, 122–4
imitatio, 42, 43
impact, 106
impetus theory, 10, 89
L'impiété des déistes (Mersenne), 133
inertia
    Descartes, 91, 103
    Galileo, 89, 90, 91
    Newton, 10, 91, 204, 217
infinity, 104, 108–21, 163, 186–91
De l'infinito, universo e mondi (Bruno),
    62, 108–9
*Initia et specimina scientiae novae
    generalis pro instauratione et
    augmentis scientiarum ad publicam
    felicitatem* (Leibniz), 39
insects, 51
instantaneous velocity, 189
instinct, 132
Institutiones rei herbarie (Tournefort),
    179
instruments, 183–6 *see also*
    microscope; telescope
insulators, 148
integration, 189
intellectuals, 4
intervention, 183
Introduction to true physics (Keill),
    173
inventions, 24, 41
Isagoges breves in anatomiam, 45
Italy, universities, 192–3

De Jesu Christi Salvatoris nostri vero anno natalitio (Kepler), 68
Jesuits, 50, 65, 77, 86, 93, 151–3
Jesus Christ, 18, 22, 133, 168, 225, 226
John Paul II, Pope, 94–5
Johnson, F. J., 199
Johnston, John, 181
Joshua, 60, 79
Journael (Beeckman), 100
Journal des Savants, 202
Jupiter, 49, 65, 67, 75, 112, 207

Kant, Immanuel, 164
Keill, John, 173–4, 214
Kepler, Johannes, 1, 12, 28, 40, 224
  Ad Vitelionem paralipomena, 68
  Astronomia nova seu Physica coelestis, 3, 68, 69–70, 71, 87
  and Bergerac, 117
  and Borel, 117
  and Brahe, 64–5, 67, 68, 69
  and Bruno, 111
  celestial kinematics, 87
  Copernicanism, 56, 61, 63, 66–7, 69, 81
  The Cosmographical Mystery, 61, 66, 67–8, 70, 71
  and Descartes, 72, 101, 106
  Dioptrice, 70, 210
  Dissertatio cum Nuncio Sidereo, 70, 111
  The Dream, 115–16, 121
  Earth, 67, 68, 111, 112
  Epitome of Copernican Astronomy, 70
  Five Books of the Harmonies of the World, 2, 70
  and Fludd, 142
  De fundamentis astrologiae certioribus, 68
  and Galileo, 50, 67–8, 72, 74, 92–3, 111–12, 113
  and Hariot, 54
  Hermeticism, 26, 71
  and Huygens, 118–19
  infinity, 110–12, 115, 187

De Jesu Christi Salvatoris nostri vero anno natalitio, 68
  law of areas, 205
  magnetism, 149, 152
  mechanical philosophy, 125–6
  moon, 69, 115–16
  multiplicity of worlds, 110–12
  Mysterium cosmographicum, 61, 66, 67–8, 70, 71
  New Astronomy, 44
  and Newton, 72, 213
  planetary motion, 69, 70–2, 205
  planets, 66–72, 75–6, 112, 186
  and Ptolemy, 81
  Somnium seu opus posthumum de astronomia lunari, 115–16, 121
  stars, 111, 112, 118–19
  De stella nova, 68
  sun, 67, 68, 69, 70, 71–2, 111, 112, 118, 119, 149, 152
Keynes, John Maynard, 215
Keyser, Konrad, 31
kinematics, 137
kinetic energy, 136
kinetics, 124
Kircher, Athanasius, 151–2, 153, 155, 156, 168
Kline, M., 187, 189, 191
knowledge, 18–28, 37–8
Komenski, Johannes Amos *see* Comenius, Amos
Königliche Preussische Akademie der Wissenschaften, 201
Koyré, Alexandre
  Cartesian laws of nature, 106, 107
  and Galileo, 113
  and Kepler, 71, 111
  metaphysical principles, 9
  Newtonian physics, 204
Kraft, F., 200
Kubrin, David, 219, 220
Kuhn, Thomas, 9, 60, 109

la Mothe le Vayer, Françoise de, 117
language botanical/zoological classification, 177–9
  historical continuity, 9

Leibniz, 190
  mechanical philosophy, 129
  secret knowledge, 20, 21
latitude, 150
Laudan, H., 122
Lavoisier, Antoine Laurent, 147
law, 192
Le Goff, J., 4, 6
Lectiones opticae (Barrow), 210
Lectiones Opticae (Newton), 213
Leeuwenhoeck, Antony van
  letters to Royal Society, 161, 162
  microscopes, 51, 52, 53, 160, 161
  spermatozoa, 162
  university study, 193
Leibniz, Gottfried Wilhelm, 1, 25, 26,
  165
  academies, 201
  calculus, 188, 189–91, 204, 214
  critique of mechanics, 135–8
  and Descartes, 99, 135, 136
  dynamics, 136, 137
  Earth, 171–3
  Essay de dynamique, 137
  God, 135–6, 138, 171, 218
  *Initia et specimina scientiae novae
    generalis pro instauratione et
    augmentis scientiarum ad publicam
    felicitatem*, 39
  mathematics, 12, 102
  matter, 135, 136, 137
  nature, 40, 138
  and Newton, 188, 189–90, 204, 209,
    214, 218
  *Nouveaux essais sur l'entendement
    humain*, 137
  physics, 12, 102, 135–7
  Protogaea, 168, 171–3
  and Schott, 153
  Societas Regia Scientiarum, 201
  Specimen dynamicum, 137
  statics, 136, 137
  universal Christianity, 40
  universal language, 177
  universities, 201
Leighton, R. B., 190, 212
Leméry, Nicholas, 146

Leo X, Pope, 34
Leonardo di Capua, 99, 197
Leonardo da Vinci, 31, 33–5, 44–5,
  166
Leopold, Prince, 196
Leopoldinisch-Carolinische Deutsche
    Akademie der Naturforscher, 200
leprosy, 27
Letter on Sunspots (Galileo), 78–9
Letter to Castelli (Galileo), 78, 79, 81
Letter to Liceti (Galileo), 113
Letter to Madame Christina of Lorraine
    (Galileo), 82, 202
*Letters of the Carmelite Paolo Antonio
    Foscarini on the opinions of
    Pythagoras and Copernicus*, 81
Lettres philosophiques (Voltaire), 106,
    209, 221–2
Leviathan (Hobbes), 125, 193
Lhwyd, Edward, 168, 169
Liber de mineralibus (Paracelsus), 141
libertine movement, 132–3, 220, 221,
    226
Liceti, Fortunio, 98, 113
The Life of Descartes (Baillet), 3
light, 104–5, 123, 145, 210–12
light bulb, 9–10
Linnaeus, 175, 176, 179, 180–1, 182
Lippi, Filippo, 32
Lister, Martin, 168, 169
liver, 157
Locke, John, 99, 125, 137
lodestone, 148, 150
logic, 15
Lomazzo, Paolo, 31
Lombardo Radice, L., 187
London, 199–200
longitude, 37
Lorini, Bonaiuto, 36–7
Lorini, Nicholas, 77, 79, 82
Louis XII of France, 34
Louis XIV of France, 198–9
love, 20
Lovejoy, Arthur O., 110
Lower, Richard, 160
Lower, William, 49
Lucian, 115

Lucretius, 27, 114
and Borel, 117
and Bruno, 110, 184
infinite void, 109
mechanical philosophy, 132
nature, 133
De rerum natura, 117, 184
and Scilla, 168
Lullism, 26
Luporini, C., 35
Luria, S. E., 176
Luther, Martin, 60

McGuire, J. E., 224
Mach, Ernst, 92
Machiavelli, Niccolo, 142
Machinae novae (Veranzio), 31
McLuhan, M., 41
macrocosm, 20, 126, 141, 143, 216–17
Maestlin, Michael, 61, 66, 67
Magalotti, Lorenzo, 153, 196
Magellan, Ferdinand, 49, 53
Magia adamica (Vaughan), 21
Magia naturalis (Porta), 151
magic, 19–22, 62
Aristotle, 19
Bacon, 24
Corpus Hermeticum, 19
Gilbert, 150
and Jesuits, 151–3
Kircher, 156
Mersenne, 133, 142
Paracelsus, 140
Magini, Giovanni Antonio, 50
*Magnes sive de arte magnetica opus
tripartitum* (Kircher), 152
magnesium sulfate, 144
De magnete (Gilbert), 115, 149–50
magnetism, 26, 29–30, 122, 124, 139,
148–56
magnetite, 148
Maldonado, T., 23
Malebranche, Nicholas, 163, 201
Malpighi, Marcello, 1, 51, 52, 126,
160, 182, 202
Mamiani, Maurizio, 215, 227, 228
The Man in the Moon (Godwin), 116

Manetti, Giannozzo, 42
Manuel, Frank, 214, 221, 222, 227,
228
Manuzio, Aldo, 42
Manzoni, Alessandro, 3
Maria Celeste, Sister, 94
Mars, 65, 67, 68, 112
Marsili, Ferdinand, 196, 202
Martin, H. J., 42
Martin, Jean, 31
Martini, Francesco di Giorgio, 31
mass, 10, 204
material strengths, 95–6
materialism, 132, 135, 138
mathematics, 139
Alberti, 32
Descartes, 102–3, 107, 188
Galileo, 12, 85, 102
Huygens, 12, 102
Leibniz, 12, 102
and magnetism, 149
Newton, 12, 102, 188–9, 203
Pascal, 12, 102
as theoretical tool, 186–91
university curricula, 192–3
Mathias, P., 199
matter
Boyle, 144–5
Descartes, 103–5, 133, 134, 135, 184
Galileo, 83–4
Hooke, 123
Leibniz, 135, 136, 137
Newton, 204, 217
Paracelsus, 141
reality, 125
Matthew, Gospel according to, 18
Maupertuis, Pierre-Louis Moreau de,
201
Maurice of Nassau, Prince, 100
mechanical philosophy, 6, 122–38, 209
animals, 129–32
Boyle, 126, 133, 134, 144–5
and chemistry, 144–8
Descartes, 123, 126, 129–30, 137,
209
dynamics, 122, 124
God, 132–5

Hermeticism, 132, 133
Kepler, 125–6
nature, 122–3, 125, 126, 127–8
Newton, 134–5, 225
Mechaniche (Galileo), 95
Mechanicorum libri (Monte), 16, 31
mechanics, 15–17, 139
    Descartes, 102
    Dijksterhuis, 124
    Galileo, 124
    Leibniz, 135–8
    mechanical philosophy, 124–7
    Newton, 124
    secrecy, 23
    and vitalism, 145–7
    see also engineers
mechanization, 124
Medici, Giuliano de, 75
medicine
    Aristotle, 19
    Borel, 43–4
    and chemistry, 140
    Diderot, 132
    Malpighi, 126
    Paracelsus, 44, 140–1, 142
    university curricula, 192
    Vesalius, 30, 46
Meditationes de prima philosophia
    (Descartes), 101, 127
Melanchthon, Philip, 36, 60, 115,
    200
memory, 180–1
men
    mechanical philosophy, 129–32
    New World, 54–5
mercury (chemical element), 141, 145
Mercury (planet), 65, 112
Mersenne, Marin
    alchemy, 24
    Copernicanism, 62–3
    correspondences, 197
    and Descartes, 72, 94, 106
    and Fludd, 142
    and Galileo, 74
    and Huygens, 107
    magic, 133, 142
    man's abilities, 3

objective/subjective qualities of
    bodies, 125
*Questionnes celeberrimae in
    Genesim*, 142
truths in physics, 128
Vérité des sciences, 84
metallurgy, 16, 36
metaphor, 21, 122
metaphysical skepticism, 127
metaphysics, 9, 135–7
Météores (Descartes), 101, 104–5
Meteorologica (Aristotle), 166
meteorological phenomenon, 124
Methodus Plantarum nova (Ray), 178
micro-earth, 150
microbiology, 51
microcosm, 20, 126, 141, 143, 216–17
Micrographia (Hooke), 123, 166, 210
microscope, 51–3, 137, 160, 162, 186
Microscopia (Hooke), 52
microscopium, 51
Milky Way, 48
millenarianism, 21, 173, 227
Milton, John, 2, 94, 193
De Mineralibus, 166
mineralogy, 36
mining technology, 16, 36
Minotaur, 40
miracles, 133
modernity, 40
Mohammed, 133
Molière, 2
Mona Lisa (Leonardo da Vinci), 33–4
monads, 136–7
Monas hieroglyphica (Dee), 62
Mondino de' Luzzi, 45
monotremes, 162
Montaigne, Michel Eyquem de, 55, 117
Monte, Guidobaldo del, 16, 31, 73
Monteverdi, Claudio, 2
moon, 74–5
    Aristotelian philosophy, 13
    Brahe, 65
    Burnet, 169–70
    Copernicus, 57
    Galileo, 48–9, 92–3
    Kepler, 69, 115–16

moon (*cont.*)
  Newton, 208
More, Henry, 113–14, 115, 134, 213
More, Thomas, 30
Moses, 133, 135, 168, 221, 222, 226
motion
  ancient physics, 11
  Aristotelian philosophy, 12–13,
    73–4, 89, 90, 184
  Bruno, 108
  Copernicus, 103
  Descartes, 11–12, 91, 103, 104–6,
    133, 136
  Galileo, 73–4, 83–4, 88–93, 95,
    96–8, 103, 106
  Hobbes, 125
  Leibniz, 136
  mechanical philosophy, 122
  Newton, 10, 91, 204–5, 206, 217
  reality, 125
  terrestrial world, 12–13
  violent versus natural, 12, 14
motion, planetary
  Aristotelian philosophy, 13, 159
  Bruno, 109
  Copernicus, 56, 57–60
  Descartes, 105–6
  Galileo, 87–8, 90–1
  Kepler, 69, 70–2, 205
  Newton, 134, 207, 208, 218
  Plato, 90–1
  Ptolemy, 13–14, 57, 59
De motu cordis (Harvey), 26, 158
De motu corporum ex percussione
    (Huygens), 106
De motu corporum in gyrum (Newton),
    213
De Motu (Galileo), 73, 95, 96
*De mulierum organis generationi
    inservientibus* (Graaf), 162
De mundi systemate (Newton), 223
*De mundo nostro sublunari
    philosophia nova* (Gilbert), 44
Mundus subterraneus (Kircher), 168
Muraro, L., 151
muscles, 130, 131, 160
Muses, 168

music, 15, 155–6
Mysterium cosmographicum (Kepler),
    61, 66, 67–8, 70, 71
Mysterium Magnum, 141
mysticism, 20, 26, 132, 133, 141

Narratio Prima (Rheticus), 58, 61
Natio Germanorum, 56
De natura fossilium (Agricola), 36
Natural Parts of Medicine (Fernel),
    159–60
nature, 5, 19, 50–1
  and art, 127–8
  Bacon, 35, 40, 166, 183–4
  Boyle, 40, 133, 134
  Descartes, 40, 103, 106–7, 123, 134,
    135
  Epicureans, 133
  Galileo, 40, 78–9, 84–5, 106
  Gassendi, 127
  Helmont, 143
  Leibniz, 40, 138
  Lucretians, 133
  and man, 40
  mechanical philosophy, 122–3, 125,
    126, 127–8
  Newton, 206, 219, 227–8
  and time, 6, 164–74
La nature et diversité des poissons
    (Belon), 4
Neapolitan Accademia degli
    Investiganti, 197
neoplatonic mysticism, 14, 141
Neoplatonism, 26, 71, 117
Neopythagorism, 71
Netherlands, universities, 193–4
New Astronomy (Kepler), 44
New Atlantis (Bacon), 51
*The New Attractive, Containing a Short
    Discourse of the Magnet or
    Lodestone* (Norman), 3, 29–30,
    150
New Experiments (Boyle), 210
New Science (Vico), 129, 222
A New Theory of the Earth (Whiston),
    170–1
New World, 53–5

Newton Historian (Manuel), 222
Newton, Humphrey, 214
Newton, Isaac, 1, 5, 40, 120, 165,
    203–29
  alchemy, 7, 26, 215, 224–5
  and Bacon, 107
  Bible, 220–1, 225, 226–8
  and Boyle, 107, 213
  Boyle Lectures, 173
  calculus, 7, 188–91, 203, 204, 213
  chronology, 220–2
  *Chronology of Ancient Kingdoms
    Amended*, 220
  cosmic cycles, 218–20
  and Descartes, 106, 107, 134, 189,
    203, 206, 213
  Earth, 207, 219
  and Euclid, 203, 204
  force, 10, 204, 205
  "General Scholium", 134, 204–5,
    208–9, 224
  geometry, 188–9, 203
  God, 208, 217, 218, 225–7, 228
  gravity, 10, 204, 207–9, 213, 217,
    223
  and Hooke, 107, 213, 214
  hypotheses, 131, 209, 219
  inertia, 10, 91, 204, 217
  and Kepler, 72, 213
  knowledge of the ancients, 222–4
  Lectiones Opticae, 213
  and Leibniz, 188, 189–90, 204, 209,
    214, 218
  life, 212–14
  mathematics, 12, 102, 188–9, 203
  matter, 204, 217
  mechanical philosophy, 134–5, 225
  mechanics, 124
  micro-entities, 122
  motion, 10, 91, 204–5, 206, 217
  De motu corporum in gyrum, 213
  De mundi systemate, 223
  nature, 206, 219, 227–8
  Opera omnia, 215
  Opticks, 2, 134, 210–12, 213–14,
    216–17, 228
  The Original of Monarchies, 221
  *Philosophiae naturalis principia
    mathematica*, 107
  see also *Principia*
  physics, 10, 12, 14, 102, 203–12,
    216–17
  Principia, 10, 188–9, 203–8, 213,
    215, 220, 224, 227
  De quadratura curvarum, 189
  "Queries," 216–17
  Royal Society, 149, 212, 214, 215
  scientific instruments, 183
  Scolii classici, 215
  space, 26, 204–5
  time, 204–5, 218–19
  Tractatus de quadratura curvarum,
    189
  The Treatise on the Apocalypse, 215,
    228
  triboelectricity, 149
  voids, 137
Newtonians
  Burnet's sacred theory of the Earth,
    170
  opposition to "world-makers,"
    173–4
Nicholas of Cusa, 110, 112, 115, 118
Nicolson, M., 115
nitric acid, 144
Noah, 222
Nocenti, L., 152
Norman, Robert, 3, 29–30, 150
De nostri temporis studiorum ratione
    (Vico), 128–9
*Nouveaux essais sur l'entendement
    humain* (Leibniz), 137
Nova de universis philosphia (Patrizi),
    3, 44, 80
Novara, Domenico Maria da, 56
Novarum observationum libri
    (Roberval), 63
Novo teatro di machine et edificii
    (Zonca), 31, 44
Novum Organum (Bacon), 3, 44, 227
Numa Pompilius, 223

occult, 62, 99, 127, 139
Ockham, William of, 23, 206

Ockham's razor, 206
Odierna, Giambattista, 51
Olber, Heinrich, 112
Oldenburg, Henry, 200, 202, 211, 219, 225
Olmi, G., 196
On Fortifications (Galileo), 74
On the Infinite Universe and Worlds (Bruno), 62, 108–9
On Mechanics (Galileo), 74
On the Most Recent Phenomena of the Aetherial World (Brahe), 65
On the Motion of Animals (Borelli), 131
On the New Star (Brahe), 64
On the Use of Parts (Galen), 44
Opera omnia (Newton), 215
Opere (Galileo), 202
Opticks (Newton), 2, 134, 210–12, 213–14, 216–17, 228
optics, 122
Order of the Rosicrucians, 100, 142
Oresme, Nicole, 5
organicism, 20
Origen, 22
The Original of Monarchies (Newton), 221
orogeny, 168
Orpheus, 22, 62
Orsini, Cardinal Alessandro, 82
De ortu et causis subterraneorum (Agricola), 36
Ortus medicinae (Helmont), 143
Osiander, Andreas, 81, 108
Oviedo y Valdes, Gonzalo Fernández de, 53–4
oviparous animals, 162
ovism, 26, 160–3
oxidation, 147
oxygen, discovery of, 10
oxygenation, 160

Padua, 47
paganism, 222
Pagel, W., 158, 159
Pagnoni, Silvestro, 76
painting, 32, 44

Palingenius, Stellatus, 62, 115, 117
Palissy, Bernard, 29, 30, 166
Panofsky, E., 44
Pansophiae prodromus, 25
Pappus, 42
Paracelsism, 139, 140, 142–3, 146
Paracelsus
  Archidoxis, 141
  color, 211
  combustive element, 147
  Liber de mineralibus, 141
  mathematical digressions, 27
  medicine, 44, 140–1, 142
  nationality, 1
  New World, 55
  Philosophia ad Athenienses, 141
Parallel between the University of Bologna and others beyond the mountains (Marsili), 202
Paré, Ambroise, 38
Paris, 197–9
Parkinson, John, 182
particle theory, 84
Partington, J. R., 147
Parva naturalia (Aristotle), 192
Pascal, Blaise
  antiquity, 43
  Experiences nouvelles touchant le vide, 185
  mathematics, 12, 102
  nationality, 1
  objective/subjective qualities of bodies, 125
  religiosity, 120
  Traite sur l'equilibre des liqueurs, 185
  voids/atmospheric pressure, 185
Pasiphaë, 40
patents, 23
Patrizi da Cherso, Francesco, 27, 61–2, 65, 69, 71, 153
  condemnation of, 80
  Nova de universis philosphia, 3, 44, 80
Paul III, Pope, 60
Paul V, Pope, 77, 82
Pauw, Corneille de, 55
percussion, 106

Perfit Description of caelestiall Orbes (Digges), 62
Périer, Florin, 185
Peripatetics, 184, 213
perpetual continuation, 91
Perrone Compagni, V., 22
Persio tradotto (Stelluti), 51
Peter the Great, 161
Philip II of Spain, 46
Philolaus, 56, 223
Philosophia ad Athenienses (Paracelsus), 141
Philosophia Magnetica (Cabeo), 151
Philosophiae naturalis principia mathematica (Newton), 107
  see also Principia
Philosophical and Astronomical Balance (Grassi), 83
Philosophical Letters (Voltaire), 106, 209, 221–2
Philosophical Transactions, 161, 171, 202
Philosophie der Geschichte (Hegel), 55
phlogiston, 147
Physica subterranea (Becher), 146
Physico-mathesis de limine, coloribus et iride (Grimaldi), 210
physics, 10–12, 43–4
  Descartes, 11–12, 102–7, 130, 137, 203, 204, 209
  Galileo, 10, 12, 95–8
  Hermetic–Paracelsian tradition, 140
  Leibniz, 12, 102, 135–7
  Newton, 10, 12, 14, 102, 203–12, 216–17
  see also mechanics
Physics (Aristotle), 184, 192
physiognomy, Aristotle, 19
physiology, 45–6, 126, 129–31, 157–63
Picard, Jean, 198
Picatrix, 22
Piccolomini, Archbishop, 94
Pickering, A., 183
Piero della Francesca, 31
Pinax theatri botanici (Bauhin), 180
Pirotechnia (Biringuccio), 23, 31, 35–6
Pisa, 47

De piscibus marinis (Rondelet), 47
planets
  Aristotelian philosophy, 13
  Brahe, 64–5
  Bruno, 62, 109
  Copernicus, 56, 57–60
  Descartes, 105–6
  Galileo, 49, 75–6
  Kepler, 66–72, 75–6, 112, 186
  life on, 115–20
  Ptolemy, 13–14, 57, 59
  revolutionary theses, 110
  see also motion, planetary
plant classification, 175–82
plants, New World, 53–4
Plato, 15, 19, 22, 90–1, 223
Platonism, 85–6, 115, 166
Plotinus, 22
pneumatic pump, 132, 186
Poli, C., 198
political renewal, 21
politics, 127
Poliziano, Angelo, 42
Pollaiuolo, 32
Pomponazzi, Pietro, 133
Pope, Alexander, 229
Poppi, Antonio, 76, 77
Porphyry, 22
Porta, Giambattista della, 27, 80, 150, 195
Posterior Analytics (Aristotle), 192
potential, 149
Power, Henry, 50–1
practice, and discourse, 29–30, 36
precision watch, 186
preformation theory, 162–3
Principia (Newton), 10, 188–9, 203–8, 213, 215, 220, 224, 227
Principles (Descartes), 11–12, 101, 106, 107, 113, 134, 135, 153
Principles of Human Knowledge (Berkeley), 190
printing, 41–2
Proclus, 14
Prodromo ovvero saggio di alcune invenzioni nuove premesso all'Arte Maestra (Terzi), 152

Prodromus (Steno), 167
Prognostication Everlasting (Digges),
    62
projectile motion, 95, 98, 102–3
propagation, 123
Protogaea (Leibniz), 168, 171–3
protozoa, 53, 161
Providence, 135
Prutenic Tables (Rheinhold), 59, 61, 68
Psalms, Book of, 79–80
psychology, 126
Ptolemy, 44, 63, 65, 66, 139
    Almagest, 13, 42, 56, 59, 74
    and Galileo, 74
    Geography, 42
    and Kepler, 81
    planetary motion, 13–14, 57, 59
public knowledge, 22–5, 28
Pucci, Francesco, 80
De pulmonibus (Malpighi), 52, 126
pumps, 132, 186
Puritan movement, 193
Pythagoras, 19, 22, 56, 75
Pythagorean number mysticism, 142

De quadratura curvarum (Newton),
    189
De quadrupedis (Johnston), 181
quantification, 143
Quercetanus, 142
Query 31 (Newton), 134
Questionnes celeberrimae in Genesim
    (Mersenne), 142
Quondam, A., 195

Ramelli, Agostino, 31, 37–8
Ramus, Petrus, 44, 193
rarefaction, 124
Ratio ponderum librae et simbellae
    (Grassi), 84
Rattansi, P. M., 199, 224
Ray, John, 168, 169, 170, 178–9, 180
Raymond, Brother, 23
De re metallica (Agricola), 16, 23, 36
reading, 41
reality, 125
reason, 24, 99, 129

Recherche de la vérité (Malebranche),
    163
Recorde, Robert, 61, 62
Redi, Francesco, 161, 162, 196
reflection, 123
reflex action, 130
refraction, 101, 123
Regulae ad directionem ingenii
    (Descartes), 100, 223
relativity, Galileo, 88–90
Rembrandt, 2, 45
Renaissance Naturalism, 132, 133
Renaudot, Théophraste, 197
representation, 183
Representing and Intervening
    (Hacking), 183
repulsion, 148, 151, 154
Rerum medicarum Novae Hispaniae
    thesaurus, 54
De rerum natura (Lucretius), 117, 184
De rerum natura (Telesio), 80
research institutes *see* academies
respiration, 145, 147
rest *see* statics
revelations, 21
De revolutionibus orbium coelestium
    (Copernicus), 2, 56, 57–9, 60–1,
    81, 108
Rheinhold, Erasmus, 59, 61
Rheticus, Georg Joachim, 57–8, 61
rhetoric, 15
Richelet, César-Pierre, 16
Richelieu, Cardinal, 198
Richer, Jean, 198
riddles, 22
Riverius, Stephanus, 45
Rivius, Walter, 31
Roberval, Gilles Personne de, 63, 184
rocks, 165–6, 172
Roemer, Olaus, 199
Roger, J., 160
Rohault, Jacques, 154
Roman Curia, 60
Romantics, 55
Rondelet, Guillaume, 47
Rosen, E., 115
Rosenkreutz, Christian, 100

Rosicrucians, 100, 142
Rossi, P., 5, 30, 165, 178, 180, 198, 220
Rothmann, Christoph, 61, 62
Rousseau, Jean-Jacques, 4
Royal Prussian Academy of Sciences, 201
Royal Society, 24, 25, 199–200
  expulsion of Woodward, 170
  Huygens, 107
  Kircher, 152
  Leeuwenhoek's letters, 161, 162
  Malpighi's membership, 160
  and Newton, 149, 212, 214, 215
  Philosophical Transactions, 161, 171, 202, 212
  Scilla's work, 168
Rudolph II, 70
Rudolphine Tables, 68
Rudwick, M. J. S., 165
Ruini, Carlo, 47
Rules for the Direction of the Mind (Descartes), 100, 223

Saads, M., 190, 212
Saggi di naturali esperienze (Magalotti), 153, 196–7
Sagredo, Giovan Francesco, 86, 88, 92, 95, 98
saltpeter, 144
salts, 141, 144, 145
salvation, 21
Salviati, Filippo, 84, 85, 86–7, 88, 89, 91, 95, 187
Saturn, 65, 67, 75, 112, 119, 207
scabies, 27
The Sceptical Chymist (Boyle), 144
Schmitt, B., 192, 193
Schmitt, C., 85
Scholasticism, 5, 14, 42, 60, 99, 133, 209
Schott, Kaspar, 153
science fiction, 116
Scienza Nuova (Vico), 129, 222
Scilla, Agostino, 167–8
Scolii classici (Newton), 215
Scripture *see* Bible

secrecy, 18–28
Secreta secretorum (Aristotle), 18–19
semantics, 21
Seneca, 85
sense perception, 83–4, 123, 125, 126, 206–7
sensory aids, 183–6
Sépulveda, Juan Ginés de, 55
Severinus, 27, 142
Sforza, Ludovico, 33, 44
Shaffer, S., 186
Shapin, S., 186
Shea, William, 91, 97, 102, 154
Sidereus nuncius (Galileo), 48, 49, 74, 111
Simplicius, 81, 84, 85, 87, 88, 93, 98, 184
sine, law of, 101
Singer, S., 176
Singh, J., 189
Siris, 190
Sixteen Books on Plants (Cesalpino), 181
Skeptics, 84
slaves, 15, 16
Societas Regia Scientiarum, 201
Socrates, 19, 117, 226
sodium sulfate, 144
solidarity, 28
Solinas, G., 160, 172
Solmi, E., 33
Somnium seu opus posthumum de astronomia lunari (Kepler), 115–16, 121
Sophocles, 147
Sorbière, Samuel, 198
sorcery, 20, 139
Sørenson, Peter, 27, 142
soul, 19, 151
  Descartes, 129, 130
  Kepler, 69, 71–2
sound, 123
space
  Descartes, 104
  Leibniz, 137
  Newton, 26, 204–5
space, Cassirer, 102–3

Spagyric principles, 142
Specimen dynamicum (Leibniz), 137
speculation, 183
speech, 129
spermatozoa, 53, 162
Spina, Bartholomew, 60
Spinoza, Baruch, 99, 132, 135, 138
spontaneous generation, 161–2
Sprat, Thomas, 24, 25, 200
Stahl, Georg, 145–7
Starry Messenger
    see *Sidereus nuncius*
stars, 20
    Aristotelian philosophy, 13
    Bergerac, 117
    Brahe, 64, 65, 75
    Bruno, 109, 119
    Descartes, 105
    Galileo, 48, 49, 113
    Huygens, 118
    Kepler, 111, 112, 118–19
    Newton, 208
    Patrizi, 61–2
    revolutionary theses, 110
statics
    ancient physics, 11
    Descartes, 11, 103
    Galileo, 89, 124
    Leibniz, 136, 137
    mechanical philosophy, 122
    Newton, 124, 204–5
Steensen, Niels *see* Steno, Nicholas
Steinberg, S. H., 41
Steiner, Christoph, 101
De stella nova (Kepler), 68
Stelluti, Francesco, 51, 54
Steno, Nicholas (Steensen), 160, 167,
    168, 172, 196
Stevinus, Simon, 31, 37, 101
Stoicism, 15, 184
strength, 95–6
Suavius, Leo, 142
substance, and activity, 137
sulfur, 141, 145
sulfuric acid, 144
sun
    Aristotelian philosophy, 13

Brahe, 65
Bruno, 119
Copernicus, 57, 58–9, 82
Descartes, 105
Galileo, 79–80, 87
Kepler, 67, 68, 69, 70, 71–2, 111,
    112, 118, 119, 149, 152
Newton, 220
Patrizi, 61
sun worship, 26
sunspots, 75
Swammerdam, Jan, 51, 160
Sylvius, Jacobus, 46
symbolism, 190
De symphathia et antipathia rerum
    (Fracastoro), 27
sympathy, 27, 28, 148, 149
Syntagma philosophicum (Gassendi),
    127
Syphilis sive de morbo gallico
    (Fracastoro), 28

Table Talks (Luther), 60
Tabulae astronomiae (Halley), 171
Tabulae sex (Vesalius), 45
tarantism, 155–6
Tartaglia, Niccolo, 31, 188
taxonomy, 175–82
Taylor, F. S., 21
*Technica curiosa sive mirabilia artis
    libri XII* (Schott), 153
technology, 40
Tega, W., 202
telescope, 17, 48–50, 51, 52, 186, 212
Telesio, Bernardino, 27, 80, 99, 197
Telluris theoria sacra (Burnet), 169–70,
    171
Temple, William, 170
tension, 149
terella, 150
La terra del rimorso (De Martino),
    155
Tertullian, 22
Terzi, Francesco Lana, 152
Tesoro messicano, 54
Des Teutschlandts Wohlfahrt
    (Glauber), 144

Théâtres des instruments mathématiques et méchaniques (Besson), 31
Theatrum botanicum (Parkinson), 182
Theodotus, 22
theology, 60, 127, 192, 193
Theophrastus redivivus, 132
Theoremata circa centrum gravitatis solidorum, 73
theory, 183
thermometer, 186
De Thiende (Stevinus), 37
Thomas of Aquinas, 14, 15, 23, 81
Thorndike, Lynn, 19, 44
thought, 129, 130
tides, 92–3, 207, 208
time
    Cassirer, 102–3
    Descartes, 134
    discovery of, 164–5
    Leibniz, 137
    and nature, 6, 164–74
    Newton, 204–5, 218–19
Tiraboschi, Girolamo, 194–5
Titian, 32, 45
Toland, John, 132
Tolosani, G. M., 60
Torricelli, Evangelista, 1, 98, 155, 184–5, 186, 188
Torrini, M., 197
touch, 123
Tournefort, Joseph Pitton de, 176, 179–80, 182
Tractatus de quadratura curvarum (Newton), 189
De tradendis disciplinis (Vives), 30
Traité de la lumière (Huygens), 107
Traité des passions de l'âme (Descartes), 101
Traité sur l'equilibre des liqueurs (Pascal), 185
The Treatise on the Apocalypse (Newton), 215, 228
Treatise on Man (Descartes), 126, 129
Treatise on the Proportions of the Human Body (Dürer), 38

Treatise on the Sphere, or Cosmography (Galileo), 74
Trent, Council of, 79, 80
triboelectricity, 148–9
Trinitarianism, 226
Tyrocinium chimicum (Beguin), 143

Uccello, Paolo, 32
uniformitarianism, 165
Universal Natural History and Theory of the Heavens (Kant), 164
universities, 1, 4–5, 23, 192–4
    botanical gardens, 47
    chemistry, 140
    teaching of world systems, 63
Urban VIII, Pope, 85, 86, 93
Utriusque cosmi historia (Fludd), 142

vacuum *see* voids
Valla, Giorgio, 115
Valla, Lorenzo, 42
Vallisnieri, Antonio, 162
Valturio of Rimini, 31
van Heeck, Johannes, 195
La vana speculatione disingannata dal senso (Scilla), 167
Vasari, Giorgio, 45
Vaughan, Thomas, 21
velocity
    ancient physics, 11
    Aristotle, 10
    Descartes, 102–3, 106
    Galileo, 73, 96–8
    Newton, 10
De venarum ostiolis (d'Acquapendente), 131
venous system, 45
Venus, 65, 75, 112
Veranzio, Fausto, 31
De veritate Sacrae Scripturae (Tolosani), 60
Vérité des sciences (Mersenne), 84
Verne, Jules, 115
Veronese, Guarino, 42
versorium, 151
verumfactum, 128
Vesalius, Andreas, 30, 45–6, 52, 157

Vespucci, Amerigo, 49
Vesta, 223
Vico, Giambattista, 99, 128–9, 221, 222
Viète, François, 101, 102, 189
Villard de Honnecourt, 48
A Vindication of an Abstract of an Italian Book Concerning Marine Bodies (Wotton), 168
vinegar eels, 51
virtual innatism, 137
Vis electrica, 150
vis viva, 136
visions, 21
De vita coelitus comparanda (Ficino), 62
vitalism, 20, 99, 115, 130, 145–7
Vitruvius, 31, 36
Vives, Juan Luis, 30
Viviani, Vincenzo, 98, 184, 196
viviparous animals, 162
Voetius, Gijsbert, 101
voids, 143, 184–6
    ancient physics, 11
    Aristotle, 73–4, 104, 137, 184
    Descartes, 104, 105, 137, 184
    Galileo, 73–4
    Guericke, 154, 185
    Leibniz, 137
Volta, Alexander, 149
Voltaire, François Marie Arouet de, 106, 201, 209, 221–2
vortices, 105–6, 206

Wackher von Wackhenfeltz, Johannes Matthaeus, 111, 112
Wallenstein, Albrecht Wenzel Eusebius von, 66, 70
Wallis, John, 23, 189, 199, 213
water
    Aristotelian philosophy, 12
    Descartes, 104, 105
    Helmont, 143

Paracelsus, 141
Webster, C., 140
Weiditz, Hans, 47, 180
weight, 10
    ancient physics, 11
    Galileo, 73
    Helmont, 143
    Newton, 204
Wells, H. G., 115
Westfall, R. S.
    alchemy, 7–8
    Boyle, 145
    Descartes, 106
    Kepler, 69
    Leibniz, 136, 137
    Newton, 189, 203, 212, 213, 214, 222, 223, 225, 226, 227
    universities, 194
Whiston, William, 170–1
White, L. Jr., 4
Whiteside, D. T., 203
Wieland, W., 186
Wilkins, John, 116, 120, 177, 178–9
William IV of Hesse-Kassel, 61
William of Orange, 193
Willughby, Francis, 178
The Wisdom of God (Ray), 170
witchcraft, 2
woodcuts, 45
Woodward, John, 170, 171
workshops, 32–3
The World (Descartes), 101, 103, 134
Wotton, Sir Henry, 49–50
Wotton, William, 168
Wren, Christopher, 52

Yates, F., 180

Zodiacus vitae (Palingenius), 62
Zonca, Vittorio, 31, 44
zoological texts, 44, 47–8
Zoroaster, 21, 26, 62